Textbook of General Virology

This book is an introductory text that presents fundamental knowledge and recent advances in virology. It provides comprehensive coverage of different aspects like classification, structure, emerging viruses, cancer-causing viruses, and viral vaccines. It covers the basic biology of virus existence, evolution, and reoccurrence. It also incorporates the fundamentals of biophysical and biochemical aspects of viral replication. The book discusses important topics such as immunity to viral infections, bacteriophages, and techniques used in virology. The *Textbook of General Virology* is meant for undergraduate and postgraduate students of microbiology, immunology, genetics, and medicine.

Key Features:

- Discusses introductory and foundational knowledge of viruses for students of life sciences and medicine
- Covers the virus history, diversity of its infection strategy, and classification
- Summarizes the characteristics of different viruses on the basis of nucleic acid genome type
- Describes the biology of RNA and DNA viruses and their effect on cell growth control in animals and humans after infection
- Reviews topics like immunity to viruses and viral vaccines

Sachin Kumar is a Professor in the Department of Biosciences and Bioengineering, Indian Institute of Technology, Guwahati, India. He completed his PhD at the University of Maryland, USA, in the field of molecular virology. He holds a master's in veterinary science from the Indian Veterinary Research Institute. He has over 167 publications and five patents to his credit.

Textbook of General Virology

Edited by
Sachin Kumar

CRC Press
Taylor & Francis Group
Boca Raton London New York

CRC Press is an imprint of the
Taylor & Francis Group, an **informa** business

Designed cover image: Sachin Kumar

First edition published 2025
by CRC Press
2385 NW Executive Center Drive, Suite 320, Boca Raton, FL 33431

and by CRC Press
4 Park Square, Milton Park, Abingdon, Oxon, OX14 4RN

CRC Press is an imprint of Taylor & Francis Group, LLC

© 2025 selection and editorial matter, Sachin Kumar; individual chapters, the contributors

Reasonable efforts have been made to publish reliable data and information, but the author and publisher cannot assume responsibility for the validity of all materials or the consequences of their use. The authors and publishers have attempted to trace the copyright holders of all material reproduced in this publication and apologize to copyright holders if permission to publish in this form has not been obtained. If any copyright material has not been acknowledged, please write and let us know so we may rectify in any future reprint.

Except as permitted under U.S. Copyright Law, no part of this book may be reprinted, reproduced, transmitted, or utilized in any form by any electronic, mechanical, or other means, now known or hereafter invented, including photocopying, microfilming, and recording, or in any information storage or retrieval system, without written permission from the publishers.

For permission to photocopy or use material electronically from this work, access www.copyright.com or contact the Copyright Clearance Center, Inc. (CCC), 222 Rosewood Drive, Danvers, MA 01923, 978-750-8400. For works that are not available on CCC, please contact mpkbookspermissions@tandf.co.uk

Trademark notice: Product or corporate names may be trademarks or registered trademarks and are used only for identification and explanation without intent to infringe.

ISBN: 9781032439068 (hbk)
ISBN: 9781032436180 (pbk)
ISBN: 9781003369349 (ebk)

DOI: 10.1201/9781003369349

Typeset in Times
by Deanta Global Publishing Services, Chennai, India

Contents

Contributors vii

Chapter 1   History of Virology 1
Diwakar D. Kulkarni

Chapter 2   Viruses and Their Significance 6
Vijay Singh Bohara and Sachin Kumar

Chapter 3   Methods Used in Virology 11
Nilave Ranjan Bora and Sachin Kumar

Chapter 4   Transmission of Viruses 17
Shaurya Dumka and Sachin Kumar

Chapter 5   Virus Classification 24
Shinjini Bhattacharya, Sukanya Sonowal and Sachin Kumar

Chapter 6   Virus Structure 29
Shinjini Bhattacharya, Sukanya Sonowal and Sachin Kumar

Chapter 7   Virus Replication 35
Shinjini Bhattacharya, Sukanya Sonowal and Sachin Kumar

Chapter 8   Viral Pathogenesis 41
Aditya Singh Chauhan, Archita Sen and Sachin Kumar

Chapter 9   Double-Stranded DNA Viruses 47
Kiran Singh, Nilave Ranjan Bora and Sachin Kumar

Chapter 10 Single-Stranded DNA Viruses 57
Anusha Sairavi and Yoya Vashi

Chapter 11 Double-Stranded RNA Viruses 69
Nilave Ranjan Bora, Kiran Singh and Sachin Kumar

Chapter 12 Positive-Sense RNA Viruses 74
Kiran Singh and Sachin Kumar

vi   Contents

Chapter 13 Minus-Strand RNA Viruses 78

Kamal Shokeen and Sachin Kumar

Chapter 14 Retroviruses 91

Satyendu Nandy and Sachin Kumar

Chapter 15 DNA Viruses Containing Reverse Transcriptase 104

Sonali Sengupta and Baibaswata Nayak

Chapter 16 Bacteriophages 119

Vishnu Kumar

Chapter 17 Plant Viruses 137

Rashmi Singh, Sachin Kumar and Latha Rangan

Chapter 18 Oncolytic Viruses 146

Deepa Mehta and Sachin Kumar

Chapter 19 Emerging and Transboundary Viral Diseases 154

Nagendra Nath Barman and Lukumoni Buragohain

Chapter 20 Immunity to Virus Infection 174

Madhuri Subbiah and Haajira Beevi Habeeb Rahuman

Chapter 21 Viral Vaccines 182

Vijay Singh Bohara and Sachin Kumar

Index 189

Contributors

Nagendra Nath Barman
Department of Microbiology
College of Veterinary Science
Assam Agricultural University
Guwahati, India

Shinjini Bhattacharya
Department of Biosciences and Bioengineering
IIT Guwahati
Guwahati, India

Vijay Singh Bohara
Department of Biosciences and Bioengineering
IIT Guwahati
Guwahati, India

Nilave Ranjan Bora
Department of Biosciences and Bioengineering
IIT Guwahati
Guwahati, India

Lukumoni Buragohain
College of Veterinary Science
Assam Agricultural University
Guwahati, India

Aditya Singh Chauhan
Department of Biosciences and Bioengineering
IIT Guwahati
Guwahati, India

Shaurya Dumka
Department of Biosciences and Bioengineering
IIT Guwahati
Guwahati, India

Diwakar D. Kulkarni
ICAR-National Institute of High Security
   Animal Diseases
OIE Reference Laboratory for Avian Influenza
Bhopal, India

Vishnu Kumar
Institute of Anatomy and Cell Biology, Unit of
   Reproductive Biology
Justus-Liebig-University Giessen
Giessen, Germany

Deepa Mehta
Department of Biosciences and Bioengineering
IIT Guwahati
Guwahati, India

Satyendu Nandy
Department of Biosciences and Bioengineering
IIT Guwahati
Guwahati, India

Baibaswata Nayak
Department of Gastroenterology
AIIMS
New Delhi, India

Haajira Beevi Habeeb Rahuman
National Institute of Animal Biotechnology
Hyderabad, India

Latha Rangan
Department of Biosciences and Bioengineering
IIT Guwahati
Guwahati, India

Anusha Sairavi
Department of Molecular and Medical
   Genetics
Oregon Health and Science University
Portland, Oregon, USA

Archita Sen
Department of Biosciences and Bioengineering
IIT Guwahati
Guwahati, India

Sonali Sengupta
Department of Gastroenterology
AIIMS
New Delhi, India

Kamal Shokeen
Department of Biosciences and Bioengineering
IIT Guwahati
Guwahati, India

Kiran Singh
Department of Biosciences and Bioengineering
IIT Guwahati
Guwahati, India

Rashmi Singh
Department of Biosciences and Bioengineering
IIT Guwahati
Guwahati, India

Sukanya Sonowal
Department of Biosciences and Bioengineering
IIT Guwahati
Guwahati, India

Madhuri Subbiah
National Institute of Animal Biotechnology
Hyderabad, India

Yoya Vashi
Department of Surgery
City of Hope National Medical Center
Duarte, California, USA

1 History of Virology

Diwakar D. Kulkarni
Retired Principal Scientist, ICAR-National Institute of
High Security Animal Diseases, Bhopal, India

Viruses have probably existed in nature since the start of life, and diseases caused by them have been recorded in history long before viruses were known or discovered, but the evidence of their existence is comparatively recent. The word "virus" (from the Latin for "slimy liquid" or "poison") originated from the Greek word *ios* and was first used in English in 1599. The Greek word *ios* originated from the Sanskrit word *visha*, meaning "poison" (see *Webster's Dictionary*).[1] These smallest microbes exert significant forces on every living thing, including themselves. Viral infections and their consequences have not only altered history but also have powerful effects on the entire ecosystem. For this reason, virologists have gone to extraordinary lengths to study, understand, and eradicate these agents.[2]

The history of virology includes conceptual and technological inventions (Table 1.1). The discovery of filters that separated bacteria from other infectious material was one of the most important events that led to the identification of the first virus. The development of the electron microscope in the late 1930s can be considered a critical technological advancement in virology. Cell culture techniques allowing virus propagation outside animal models, next-generation sequencing, and metagenomic analyses are the more recent breakthroughs in the history of virology. Thus, new viruses of various shapes and sizes are being detected regularly that may have the capacity to infect humans, animals, and plants. In the 1910s, it was demonstrated that even bacteria have viruses.

Nene[1] and his associates have noted the proof of some possible cases of the existence of plant viruses in documents such as *Vrikshayurveda* (The Science of Plant Life) written by the physician Surapala (1000 AD), and Someshvardeva's (Chalukyan king, 12th century AD) *Abhilashtitartha Chintamani*, or *Manasollasa* (1131 AD).[3] In addition, the American Phytopathological Society's *Phytopathological Classic No. 7* describes the early history of plant virus studies.[4] Louis Pasteur (1822–1895), who with his swan-neck sterile sealed flasks, proved wrong the concept of spontaneous generation. His study on fermentation by different microbial agents showed that "different kinds of microbes are associated with different kinds of fermentations," and he soon extended this concept to microbes and diseases.[2] A disease of tobacco was identified in Holland in 1886. Adolf Meyer (1843), a German agricultural chemist, was appointed to investigate the cause of the tobacco disease; he named the lesions "mosaic" (hence it was popularly known as tobacco mosaic disease) and successfully attempted mechanical transmission of the disease using glass capillary tubes.[1] Later, in 1892, Dmitry Ivanovsky used filters to show that sap from a diseased tobacco plant remained infectious to healthy tobacco plants despite having been filtered. Independently, Martinus Beijerinck in 1898 called the infection a "contagium vivum fluidum" and the filtered, infectious substance a "virus."[4] He sparked a 25-year debate about whether these novel agents were liquids or particles. This conflict was resolved when d'Herelle developed the plaque assay in 1917 and when the first electron micrographs were taken of the tobacco mosaic virus (TMV) in 1939.[2] Meyer, Ivanovsky, and Beijerinck each contributed to the development of a new concept: a novel organism smaller than bacteria, an agent defined by the pore size of the Chamberland filter, that could not be seen under the light microscope, and could multiply only in living cells or tissue. In 1933, the electron microscope was invented by Ernst Ruska and Max Knoll, and in 1938 Bodo von Borries, Helmut Ruska, and Ernst Ruska published the first electron micrographs of ectromelia (mousepox)

DOI: 10.1201/9781003369349-1

TABLE 1.1

Some Important Events in the History of Virology

Year	Investigator(s)	Findings
1796	Edward Jenner	Application of cowpox virus for vaccination against smallpox
1885	Louis Pasteur, E. Roux	Development of rabies vaccine
1892–1898	D. Ivanovsky, M. Beijerinck	First demonstrations of a filterable plant virus: tobacco mosaic virus
1898	F. Loeffler, P. Frosch	First demonstration of a filterable animal virus: foot-and-mouth disease virus
1904–1908	V. Ellermann, O. Bang, H. Vallee, H. Carre	First demonstration of a retrovirus
1905	Spreull	Insect transmission of bluetongue virus
1910–1911	P. Rous	First demonstration of a tumor virus: Rous sarcoma virus
1915	F. Twort	Discovery of bacteriophages
1920–1921	H. G. Creutzfeldt and A. M. Jakob	Human neurological disorder of unknown etiology (later known as Creutzfeldt–Jakob disease [CJD])
1923–1928	A. Carrel, H. Maitland, M. Maitland	Tissue culture of embryo explants and first tissue culture cultivation of virus
1931	A. Woodruff, E. Goodpasture	Use of embryonated hen's eggs as a virus host
1938	Cuille and Chelle	Detection of neurological disorder in sheep (scrapie)
1938–1939	G. Kausche, P. Ankuch, H. Ruska	First electron micrograph of a virus (TMV)
1941	G. Hirst	Discovery of influenza virus hemagglutination, hemagglutination inhibition test
1944	O. Avery, C. MacLeod, M. McCarty	Identification of DNA as the material of inheritance
1945	T. Francis, J. Salk, G. Hirst, F. Davenport, E. Kilbourne, and others	Development of inactivated influenza vaccines
1948	Sanford, Earle, and Likely	Culture of isolated mammalian cells
1948–1955	J. Enders, T. Weller, F. Robbins, H. Eagle	Routine use of tissue culture to grow and study viruses
1953	J. Watson, F. Crick, M. Wilkins, R. Franklin	Discovery of the structure of DNA
1954	B. Sigurdsson	Development of the concept of slow viruses (Maedi-visna virus)
1956	H. Fraenkel-Conrat, B. Singer	Discovery of the infectivity of viral RNA
1957	A. Isaacs, J. Lindenmann	Discovery of interferon
1959	S. Brenner, R. Horne	Negative stain electron microscopy
1962	A. Klug, D. Caspar	Discovery of the principles of icosahedral virus structure
1967–1982	T. Alper, I. H. Pattison, J. S Griffith, Jones, Hunter, Prusiner and coworkers	Prions
1970	Temin and Baltimore	Discovery of reverse transcriptase
1973	D. Nathans	Completion of the restriction map of a viral genome (SV40)
1976	T. Diener	Discovery of viroids (infectious naked RNA molecules)
1977	World Health Organization	Eradication of smallpox
1981	Baltimore	First infectious clone of an RNA virus
1983	Montagnier, Barre-Sinoussi, and Gallo	Discovery of human immunodeficiency virus
2011	World Organization of Animal Health (OIE)	Global eradication of rinderpest

virus and vaccinia virus. It became an easy tool to understand the size and shape of the various viruses. A major advance was the development of negative-contrast electron microscopy in 1959 by Sydney Brenner and Robert Horne. Using this method, the virus particle was seen surrounded by electron-dense stains to produce a distinct negative image of the virus with remarkable resolution.[5]

History of Virology

Virologists agree that the viral universe is immense, and only a small fraction has been explored.[6] Indeed, virologists have elucidated new principles of life processes and have been leaders in promoting new directions in science. For example, many of the concepts and tools of molecular biology and cell biology have been derived from the study of viruses and their host cells.[2] Several new findings were applied in studying the structure, functions, and host-infecting capacity of the virus; e.g., cell culture procedures permitted isolation of the virus and the production of a vaccine by the early 1960s. Influenza virus was detected for the first time in wild birds in 1961, which led to the identification of waterfowl and shorebirds as the natural reservoir of influenza A viruses. The molecular era of virology began in the late 1970s. The development of the polymerase chain reaction (PCR) in 1983 had a very deep impact on virus research. The cloning of nucleic acid sequences in 1981 revolutionized the molecular techniques for virus detection and diagnostics. Identification by molecular means without isolation and/or in vitro propagation of the virus in cell culture became possible with these techniques.[5] The bacteriophage helped in defining some of the fundamental genetic principles through the study of mutations. As analytical and chemical procedures were developed, it was shown that viruses contained nucleic acids; viruses became key players in defining the role of nucleic acids as the database for life.[5]

The word "smallpox" appeared in the English lexicon in 1518 (*Webster's Dictionary*), though its occurrence was known for millennium, as a reference to smallpox in *Rigveda* (8000 BC) has been noted by Sharma.[7] In late 1975, a three-year-old girl from Bangladesh was the last person in the world to have naturally acquired smallpox (variola major). A person from Merca, Somalia, suffered in 1977 from a minor form of smallpox (variola minor) and recovered. In 1980, the World Health Organization (WHO) declared smallpox eradicated (eliminated), and no cases of naturally occurring smallpox have happened since.[8,9]

Rinderpest – also known as cattle plague – was a disease caused by the rinderpest virus, which primarily infected cattle, buffalo, and some other animals. Rinderpest was commonly reported in the Indian subcontinent, Middle East, and Africa. Watt (1891) stated that the losses resulting from the unchecked ravages of this disease in India were enormous. It was present during all seasons and infected thousands of cattle annually. Watt recorded that in 1880, about one million cattle were affected in India alone. Strict quarantine, improved hygiene, slaughter, and vaccinations were common practices in containing rinderpest. In 1994, the Food and Agriculture Organization (FAO) launched the Global Rinderpest Eradication Programme (GREP) with the goal of eradication by 2010.[10] Thanks to the program's global surveillance and vaccination efforts (a ring vaccination strategy similar to that applied to smallpox was used),[8], the last known rinderpest outbreak occurred in Kenya in 2001, with the last case being recorded in Mauritania in 2003. Over the next ten years, the GREP continued to search for rinderpest samples. Finding none, rinderpest was declared eradicated by the World Organization for Animal Health (OIE) on 25 May 2011. It is the first animal disease to be eradicated globally.[11]

Foot-and-mouth disease (FMD) has been known to exist for many centuries. It is claimed that the first report was from Italy in 1514.[12] However, *Lokopakara* (1025 AD), compiled by Chavundaraya,[13] described "boils of gum and hoof" as a distinct disease in cattle, a clear reference to foot-and-mouth disease, almost 500 years before the report from Italy.[1] In the history of virology, FMD has been the second known virus identified (after TMV in 1892) from animals, as identified by Loeffler and Frosch in 1899.[14]

The relationship between cancer and viruses was first discovered in domestic poultry by two Danish veterinary researchers, Ellerman and Bang, in 1908. They recorded that leukemia in chickens could be transmitted by a cell-free extract. Peyton Rous (1910–1911) later reported that solid tumors (sarcomas) were also transmissible by the Rous sarcoma virus (RSV), the first prototype retrovirus. He received the Nobel Prize in Medicine in 1966, many years after his discoveries.[8]

In the last two decades, several high-impact zoonotic disease outbreaks have been linked to bat-borne viruses. These include SARS coronaviruses and henipaviruses. It is being increasingly accepted that bats are potential reservoirs of a large number of known and unknown viruses, many of which could spill over into animal and human populations. Several studies have indicated that bats are exceptional in their ability to act as natural reservoirs of viruses and they are able to harbor more diverse viruses per animal species. One of the first bat-borne BSL4 agents identified was Hendra virus (HeV) in Australia in 1994.[15] Nipah virus (NiV) emerged in 1998 in Malaysia, which transmitted from bats to humans via pigs as an intermediate and amplifying host, resulting in 283 human cases and 109 deaths in Malaysia. Several NiV outbreaks have been reported in Bangladesh and India, resulting in significant human mortality.[16] The reservoir hosts of NiV have been identified as the large flying fox (*Pteropus vampyrus*) and small flying fox (*P. hypomenalus*) in Malaysia and the Indian flying fox (*P. giganteus*) in Bangladesh and India.[17]

It is a common understanding that viruses are disease-producing agents that must be controlled or eliminated. However, they have some beneficial properties too that can be exploited for useful purposes. Specifically, some viruses (e.g., baculovirus) have been engineered to express useful non-viral proteins or viral proteins for immunization purposes (e.g., poxvirus- and adenovirus-vectored vaccines). Lentiviruses have been modified for the purpose of inserting genes of interest into cells for research purposes and for use in gene therapy, as have adeno-associated viruses (which are actually parvoviruses). Bacteriophages are being considered in the context of controlling certain bacterial infections, and viruses have the hypothetical potential to be vectors that selectively target tumor cells for controlling cancers. In the broader context of the Earth's ecosystems, viruses are now viewed in a more positive sense, in that they may be a component of population control and perhaps a force in the evolution of species.[5]

In December 2019, a severe acute respiratory syndrome (SARS)-like illness was reported to the WHO Country Office in China. On January 12, 2020, Chinese scientists published the genome of the virus. On January 30, 2020, an outbreak was declared by the WHO. The disease spread very fast across several countries, the virus was named SARS-coronavirus-2, and its person-to-person transmission was confirmed. On March 11, the WHO declared COVID-19 (coronavirus disease 2019) as a pandemic.[18] Globally, as of 19 April 2023, there were over 763 million confirmed cases of COVID-19, including 6.9 million human deaths, reported to the WHO.[1]

REFERENCES

1. Nene, Y. L. 2007. *Asian Agri-History,* 11(1): 33–46. India: Asian Agri-History Foundation.
2. Enquist, L. W. and Racaniello, V. R. 2013. Virology: From Contagium Fluidum to Virome. In: *Fields Virology,* eds. David M. Knipe and Peter M. Howley., 6th ed. (Vol. 1). Wolters Kluwer Health.
3. Shamasastry, R. (Ed.) 1926. *Abhilash-titarthachintamani of Someshwara Deva. Bhudharakrida (V-I).* Government Branch Press, Mysore, India.
4. Johnson, J. (Tr.) 1942. *Phytopathological Classic No. 7.* The American Phytopathological Society, Ithaca, NY, 62 pp.
5. Coffey, L. L. 2012. Nature of Viruses. In: *Fenner and White's Medical Virology 2012.* Elsevier Inc. http://dx.doi.org/10.1016/B978-0-12-375156-0.00001-1 © 2012 Elsevier Inc.
6. Mettenleiter, T. C. 2017. "The First Virus Hunters." In: *Advances in Virus Research,* Vol. 99, eds. M. Beer and D. Höper. Burlington: Academic Press, pp. 1–16. http://dx.doi.org/10.1016/bs.aivir.2017.07.005
7. Sharma, G. S. 1991. *Rigved (In Hindi).* Sanskrit Sahitya Prakashan, New Delhi, India. 769 pp.
8. Taylor, M. W. (2014). Introduction: A Short History of Virology. In: *Viruses and Man: A History of Interactions.* Springer, Cham. https://doi.org/10.1007/978-3-319-07758-1_1
9. https://www.cdc.gov/smallpox/index.html
10. Ochmann, S. and Behrens H. 2018. How Rinderpest Was Eradicated? https://ourworldindata.org/how-rinderpest-was-eradicated#:~:text=Rinderpest%20was%20a%20so%2Dcalled,the%20disease%20eradication%20in%202011.
11. https://www.woah.org/fileadmin/Home/eng/Media_Center/docs/pdf/Disease_cards/RINDERPEST-EN.pdf
12. Griffith, Ralph T. H. 1973. *The Hymns of the Rigveda.* Motilal Banarsidas Publishers, New Delhi 110 007, India. 707 pp. http://ag.missouristate.edu/footm1.htm
13. Ayangarya, Valmiki Sreenivasa. (Tr.) 2006. *Lokopakara (For the Benefit of People).* Agri-History Bulletin No. 6. Asian Agri-History Foundation, Secunderabad 500 009, India, 134 pp.

14. Booss, J. & August M. J. 2013. Of Mice and Men: Animal Models of Viral Infection. In: *To Catch a Virus*, eds. J. Booss and M. J. August. ASM Press, Washington, DC, p. xxii.
15. Wang, L. F. and Anderson, D. E. 2019. Viruses in Bats and Potential Spillover to Animals and Humans. *Current Opinion in Virology*, 34: 79–89
16. Kulkarni, D. D., Tosh, C., Bhatia, S. and Raut, A. A. 2017. Nipah Virus Infection. Ch. 12 In: *Emerging and Re-emerging Infectious Diseases of Livestock*, ed. J. Bayry. Springer International Publishing AG, Switzerland, pp. 285–299. ISBN-978-3-319-47424-3 Print; 978-3-319-47426-7 (eBook)
17. Murray, K., Selleck, P., Hooper, P., Hyatt, A., Gould, A., Gleeson, L., Westbury, H., Hiley, L., Selvey, L., Rodwell, B., et al. 1995. A Morbillivirus That Caused Fatal Disease in Horses and Humans. *Science*, 268: 94–97
18. Lango, M. N. 2020. How Did We Get Here? Short History of COVID-19 and Other Coronavirus-related Epidemics. *Head & Neck*, 42: 1535–1538. https://doi.org/10.1002/hed.26275

2 Viruses and Their Significance

Vijay Singh Bohara and Sachin Kumar
Department of Biosciences and Bioengineering, Indian
Institute of Technology Guwahati, Guwahati, Assam, India

INTRODUCTION

The history of virology is remarkable as viral diseases have shaped the history and evolution of life on Earth. Viruses have not only influenced human history but also the entire ecosystem of our planet. Almost all living organisms are infected by viruses. Many viruses are non-pathogenic to humans and maintain the health of individual organisms, from fungi and plants to insects and humans. Susana Lopez, a virologist at the National Autonomous University of Mexico, said we live in a balance, in perfect equilibrium, and viruses are part of it. Therefore, if all viruses get wiped out suddenly, the world will become wonderful for a day, and then we all would die, says Tony Goldberg, an epidemiologist at the University of Wisconsin. If a layman is asked their opinion about viruses, the first thought that will likely come to mind is an infection, disease and something that is life-threatening. They would probably think of viruses such as HIV, dengue, Zika, SARS–coronavirus 2 and whatever they see on news related to a current outbreak. However, not all viruses are harmful to their hosts; rather, their hosts benefit from them (Roossinck 2011). With the use of modern genetic engineering tools, scientists have modified harmful viruses to kill cancer cells or as vaccines or as vehicles for vaccines. Treatment using viruses, known as virotherapy, has become a subject of rigorous research. This chapter describes some of the positive aspects of viruses and the importance of viruses in our day-to-day lives.

APPLICATION OF VIRUSES IN MEDICINE

Gene therapy is a technology used to treat several diseases such as inherited disorders, viral infections and cancers. The success of gene therapy depends on the effectiveness of the gene delivery system used to treat a defective gene. Viruses can be used as gene delivery systems because of their ability to inject DNA into host cells. The genetic material of a virus (DNA or RNA) can be substituted using genetic engineering tools to contain therapeutic DNA or RNA. Viral vectors can be categorized into two categories. One is integrative, integrating their genome into host chromatin, such as lentivirus and retrovirus, and the other is non-integrative, where their genome remains in the nucleus of the host cell as episomes, such as adeno-associated virus, adenovirus and herpes virus. The choice of viral vectors depends on the efficiency of foreign gene expression, safety, toxicity and stability. Scientists have used viral vectors for the treatment of several diseases like cystic fibrosis, haemophilia, muscular dystrophy and sickle cell anaemia (Ghosh et al. 2020).

Cancer still poses a challenge to human health as the survival chances of a patient suffering from advanced cancer are still low. Many viruses have been used to treat cancers. Such viruses are called oncolytic viruses as they selectively kill cancer cells. One of the major problems with oncolytic viruses as therapeutic agents against cancer is their immune rejection. Scientists have also used viruses as immunostimulatory agents in combination with radio- or chemotherapy (Alemany 2013). Many oncolytic viruses are currently in pre-clinical trials; among them, herpesvirus, adenovirus and vaccinia virus have yielded promising results.

Viruses and Their Significance

In 1991, scientists modified the human herpes simplex 1 (HSV-1) virus by knocking out the thymidine kinase (TK) gene; infection with modified HSV-1 suppressed tumour growth in mice. In 1996, adenovirus was genetically modified and entered phase I clinical trials. Three oncolytic viruses are currently used in cancer treatment: RIGVIR, a modified Riga virus used for the treatment of melanoma; Oncorine, an attenuated adenovirus approved by the Chinese Food and Drug Administration (FDA) for the treatment of head and neck cancer; and T-VEC, a modified human HSV-1 virus approved by the US FDA for the treatment of melanoma (Cao et al. 2020).

Viruses are also used as vectors for producing antigens in the infected host. Many viruses have been developed as vaccine vectors and are currently under various stages of clinical trials. Vaccinia virus and adenovirus are the major viruses that are used as vaccine vectors, as they elicit a strong immune response in the host. In 2009, a vaccinia virus-based HIV vaccine induced moderate immunity in phase III clinical trials. Several recombinant adenoviral-based vaccines for several diseases, including HIV, influenza and solid tumours, are also under various stages of clinical trials (Ura, Okuda, and Shimada 2014).

APPLICATION OF VIRUSES IN AGRICULTURE

Viruses infect plants, resulting in a significant loss in yield and quality of agricultural products. This leads to severe economic losses for farmers and the country as well. Recent studies have shown that viruses can also be advantageous to plants. Viruses help plants overcome extreme environmental conditions such as drought, cold and extreme temperatures. Viruses reduce the need for chemical fertilizers and pesticides, thereby enhancing their value and growth potential. Some viruses enhance the beauty of ornamental plants, such as the tulip breaking virus. Viruses such as the brome mosaic virus, cucumber mosaic virus and tobacco mosaic virus help plants to overcome drought and extreme cold conditions. Some persistent viruses, such as the white clover cryptic virus, prevent nodulation in legumes in the presence of sufficient nitrogen. Viruses that result in mild symptoms upon infection in plants are being used to confer cross-protection against other severe viruses. The notion that viruses are always harmful to plants may not be correct, as several persistent viruses belonging to families such as *Chrysoviridae*, *Endornaviridae*, *Partitiviridae* and *Totiviridae* have developed long-term relationships with their hosts. These are common in crops such as peppers, rice, beans, carrots, figs, radishes, melons, barley and avocados. The beneficial effects of viruses have still not been fully studied by scientists. As the human population is rapidly increasing, it has become imperative for horticulturists, crop scientists and plant pathologists to look for environmentally friendly ways to enhance crop production. Viruses do hold strong potential in this regard, as they can help in cost-effective and environmentally friendly improvements in crop production (Roossinck 2015).

ROLE OF VIRUSES IN ECOLOGY

Viruses control the microbial population by feeding on them. Viruses reprogram host cellular machinery for their benefit. The ultimate result is the multiplication of the virus and the death of the host cells. Therefore, viruses have a huge impact on the global biogeochemical cycle as they act as microbial predators. The influence of viruses on the biogeochemical cycle remains unappreciated (Roucourt and Lavigne 2009). It is presumed that most viruses, especially phages that feed on bacteria, are ecologically important; however, archaeal and eukaryotic viruses are also crucial components of most ecosystems. The global estimated count of viruses is around 1e31, and the average half-life of free viruses in most ecosystems is ~48 h, which means 1e25 microbes or about 100 million metric tons of microbes, die every minute due to viruses. Viral predation, together with grazing by protists, maintains a microbial population less than the carrying capacity of the system. Hence, microbial diversity is controlled simultaneously by predation by viruses and substrate availability (Sandaa et al. 2009).

In marine ecosystems, photosynthesis results in the production of half of the organic matter. This organic matter supports the growth of heterotrophic microbes (both Bacteria and Archaea). Viruses and protists, in turn, feed on these heterotrophic microbes. This results in the dissolution of organic matter, which can further be utilized by other heterotrophic microbes (Fuhrman and Noble 1995). Therefore, in the oceanic world, virus-mediated host mortality enhances net respiration, carbon dioxide release and nutrient recycling. Viruses also release nutrients by killing diatoms, dinoflagellates and cyanobacteria (Suttle, Chan, and Cottrell 1990).

Viruses also drive the evolution of microbes by natural selection through lateral gene transfer. Viruses also transfer genetic material from one ecosystem to another (Short and Suttle 2005; Breitbart, Miyake, and Rohwer 2004). Viruses transfer genes of ecological importance from one host to another host. For example, viruses infecting cyanobacteria such as *Prochlorococcus* and *Synechococcus* have genes encoding proteins essential in photosynthesis (Lindell et al. 2004; Mann et al. 2003). Viruses express these proteins during infection to keep host cells alive. Our knowledge of environment-friendly viruses is still limited. With the advancement of metagenomics and other genomic tools, novel viruses and their ecological and evolutionary roles can be better understood.

APPLICATION OF VIRUSES IN THE FOOD INDUSTRY

From fresh to frozen to ready-to-eat meals, our food is under constant threat of contamination by microbial pathogens. Ever thought of how a virus can be useful in the food industry? Well, certain viruses have the potential to be used as biocontrol in the food industry.

Bacteriophage, a virus that infects only specific bacteria, is one such virus. The fact that only bacteria are infected and lysed by bacteriophages while being harmless to mammalian cells makes them a potential biocontrol and bio-preservative option for food safety concerns (García et al. 2008). Felix d'Herelle discovered bacteriophages in 1917, and the utility of these "bacteria eaters" for dealing with bacterial diseases was rapidly recognized (Sulakvelidze, Alavidze, and Morris 2001). Foodborne illness is a major health concern, and traditional methods for microbial control, such as pasteurization, irradiation and chemical preservatives, can largely reduce microbial populations. However, many limitations are associated with these methods, including high cost and negative effects on the nutritional quality of food. Good bacteria also get killed along with bad bacteria when using such procedures. Hence, these shortcomings can be overcome by using bacteriophage as a green and natural method in which specific bacteria can be targeted by lytic bacteriophages isolated from the environment (Moye, Woolston, and Sulakvelidze 2018). The most common food-borne bacterial pathogens are *Listeria monocytogenes*, *Salmonella* spp., *Escherichia coli*, *Shigella* spp. and *Campylobacter jejuni*. A511 bacteriophage has been shown to reduce *Listeria monocytogenes* levels in experimentally contaminated cheese (Guenther et al. 2009). In other studies, SJ2 phage administration resulted in the reduction of *Salmonella* colonies in experimentally contaminated eggs and ground pork (Hong et al. 2016); FAHEc1 phage showed up to 4 logs of decreased contamination of *Escherichia coli* in raw and cooked beef (Hudson et al. 2015); a cocktail of three different phages (SD-11, SF-A2, SS-92) reduced *Shigella* levels by up to 4 logs on spiced chicken pieces (Zhang, Wang, and Bao 2013); and Cj6 phage significantly decreased the levels of *Campylobacter* in beef (Bigwood et al. 2008). Some examples of phage products that have been approved for food safety applications are listed in Table 2.1.

TABLE 2.1
Phage Products Approved for Food Safety Applications

Phage Product	Target Organism	Reference
EcoShield™	*E. coli* O157:H7	(Carter et al. 2012)
Listex™	*L. monocytogenes*	(Soni, Nannapaneni, and Hagens 2010)
SalmoFresh™	*Salmonella* spp.	(Sukumaran et al. 2016)
ShigaShield™	*Shigella* spp.	(Soffer et al. 2017)

REFERENCES

Alemany, R. 2013. "Viruses in cancer treatment." *Clin Transl Radiat Oncol* 15 (3):182–8. doi: 10.1007/s12094-012-0951-7.

Bigwood, T., J. A. Hudson, C. Billington, G. V. Carey-Smith, and J. A. Heinemann. 2008. "Phage inactivation of foodborne pathogens on cooked and raw meat." *Food Microbiol* 25 (2):400–6. doi: 10.1016/j.fm.2007.11.003.

Breitbart, M., J. H. Miyake, and F. Rohwer. 2004. "Global distribution of nearly identical phage-encoded DNA sequences." *FEMS Microbiol Lett* 236 (2):249–56. doi: 10.1016/j.femsle.2004.05.042.

Cao, G. D., X. B. He, Q. Sun, S. Chen, K. Wan, X. Xu, X. Feng, P. P. Li, B. Chen, and M. M. Xiong. 2020. "The oncolytic virus in cancer diagnosis and treatment." *Front Oncol* 10:1786. doi: 10.3389/fonc.2020.01786.

Carter, C. D., A. Parks, T. Abuladze, M. Li, J. Woolston, J. Magnone, A. Senecal, A. M. Kropinski, and A. Sulakvelidze. 2012. "Bacteriophage cocktail significantly reduces Escherichia coli O157: H7 contamination of lettuce and beef, but does not protect against recontamination." *Bacteriophage* 2 (3):178–185. doi: 10.4161/bact.22825.

Fuhrman, Jed A., and Rachel T. Noble. 1995. "Viruses and protists cause similar bacterial mortality in coastal seawater." *Limnol Oceanogr* 40 (7):1236–42.

García, P., B. Martínez, J. M. Obeso, and A. Rodríguez. 2008. "Bacteriophages and their application in food safety." *Lett Appl Microbiol* 47 (6):479–85. doi: 10.1111/j.1472-765X.2008.02458.x.

Ghosh, Sumit, Alex M. Brown, Chris Jenkins, and Katie Campbell. 2020. "Viral vector systems for gene therapy: A comprehensive literature review of progress and biosafety challenges." *Appl Biosaf* 25 (1):7–18. doi: 10.1177/1535676019899502.

Guenther, S., D. Huwyler, S. Richard, and M. J. Loessner. 2009. "Virulent bacteriophage for efficient biocontrol of Listeria monocytogenes in ready-to-eat foods." *Appl Environ Microbiol* 75 (1):93–100. doi: 10.1128/AEM.01711-08.

Hong, Y., K. Schmidt, D. Marks, S. Hatter, A. Marshall, L. Albino, and P. Ebner. 2016. "Treatment of salmonella-contaminated eggs and pork with a broad-spectrum, single bacteriophage: Assessment of efficacy and resistance development." *Foodborne Pathog Dis* 13 (12):679–88. doi: 10.1089/fpd.2016.2172.

Hudson, J. A., C. Billington, T. Wilson, and S. L. On. 2015. "Effect of phage and host concentration on the inactivation of Escherichia coli O157:H7 on cooked and raw beef." *Food Sci Technol Int* 21 (2):104–9. doi: 10.1177/1082013213513031.

Lindell, D., M. B. Sullivan, Z. I. Johnson, A. C. Tolonen, F. Rohwer, and S. W. Chisholm. 2004. "Transfer of photosynthesis genes to and from Prochlorococcus viruses." *Proc Natl Acad Sci USA* 101 (30):11013–18. doi: 10.1073/pnas.0401526101.

Mann, N. H., A. Cook, A. Millard, S. Bailey, and M. Clokie. 2003. "Marine ecosystems: Bacterial photosynthesis genes in a virus." *Nature* 424 (6950):741. doi: 10.1038/424741a.

Moye, Z. D., J. Woolston, and A. Sulakvelidze. 2018. "Bacteriophage applications for food production and processing." *Viruses* 10 (4). doi: 10.3390/v10040205.

Roossinck, M. J. 2011. "The good viruses: Viral mutualistic symbioses." *Nat Rev Microbiol* 9 (2):99–108. doi: 10.1038/nrmicro2491.

Roossinck, M. J. 2015. "A new look at plant viruses and their potential beneficial roles in crops." *Mol Plant Pathol* 16 (4):331–3. doi: 10.1111/mpp.12241.

Roucourt, B., and R. Lavigne. 2009. "The role of interactions between phage and bacterial proteins within the infected cell: A diverse and puzzling interactome." *Environ Microbiol* 11 (11):2789–805. doi: 10.1111/j.1462-2920.2009.02029.x.

Sandaa, R. A., L. Gomez-Consarnau, J. Pinhassi, L. Riemann, A. Malits, M. G. Weinbauer, J. M. Gasol, and T. F. Thingstad. 2009. "Viral control of bacterial biodiversity-evidence from a nutrient-enriched marine mesocosm experiment." *Environ Microbiol* 11 (10):2585–97. doi: 10.1111/j.1462-2920.2009.01983.x.

Short, C. M., and C. A. Suttle. 2005. "Nearly identical bacteriophage structural gene sequences are widely distributed in both marine and freshwater environments." *Appl Environ Microbiol* 71 (1):480–6. doi: 10.1128/AEM.71.1.480-486.2005.

Soffer, N., J. Woolston, M. Li, C. Das, and A. Sulakvelidze. 2017. "Bacteriophage preparation lytic for Shigella significantly reduces Shigella sonnei contamination in various foods." *PLoS One* 12 (3):e0175256. doi: 10.1371/journal.pone.0175256.

Soni, K. A., R. Nannapaneni, and S. Hagens. 2010. "Reduction of Listeria monocytogenes on the surface of fresh channel catfish fillets by bacteriophage Listex P100." *Foodborne Pathog Dis* 7 (4):427–34. doi: 10.1089/fpd.2009.0432.

Sukumaran, A. T., R. Nannapaneni, A. Kiess, and C. S. Sharma. 2016. "Reduction of Salmonella on chicken breast fillets stored under aerobic or modified atmosphere packaging by the application of lytic bacteriophage preparation SalmoFreshTM." *Poult Sci* 95 (3):668–75. doi: 10.3382/ps/pev332.

Sulakvelidze, A., Z. Alavidze, and J. G. Morris, Jr. 2001. "Bacteriophage therapy." *Antimicrob Agents Chemother* 45 (3):649–59. doi: 10.1128/AAC.45.3.649-659.2001.

Suttle, Curtis A., Amy M. Chan, and Matthew T. Cottrell. 1990. "Infection of phytoplankton by viruses and reduction of primary productivity." *Nature* 347 (6292):467–9.

Ura, T., K. Okuda, and M. Shimada. 2014. "Developments in Viral Vector-Based Vaccines." *Vaccines (Basel)* 2 (3):624–41. doi: 10.3390/vaccines2030624.

Zhang, H., R. Wang, and H. Bao. 2013. "Phage inactivation of foodborne Shigella on ready-to-eat spiced chicken." *Poult Sci* 92 (1):211–17. doi: 10.3382/ps.2011-02037.

3 Methods Used in Virology

Nilave Ranjan Bora and Sachin Kumar
Department of Biosciences and Bioengineering,
Indian Institute of Technology Guwahati, Guwahati, Assam, India

INTRODUCTION

The study of viruses and viral diseases employs a variety of approaches. The area is continuously evolving due to the development of new techniques and technologies. We will discuss a summary of the methods that are frequently employed in diagnostic virology.

Diagnostic virology aims to identify the virus responsible for clinical signs and symptoms. Procedures that are most commonly used are:

1) Detection of immunological response (antibody or cell-mediated) to the virus by immunological assay
2) Staining specimens or tissue sections to identify the agent (light and electron microscopy)
3) Isolation and identification of the causative agent
4) Detection of viral nucleic acid

DETECTION OF IMMUNOLOGICAL RESPONSE

Identifying a virus associated with the disease observed or assessing an individual's response to a vaccination can be challenging. In such instances, indirect measurement techniques, such as assessing antibody response to the virus of interest, are required. The most common techniques include:

- Agar gel immunodiffusion (AGID)
- Agar gel precipitin (AGP)
- Complement fixation (CF)
- Enzyme-linked immunosorbent assay (ELISA)
- Hemagglutination assay
- Hemagglutination inhibition (HI)
- Indirect fluorescent antibody (IFA)
- Latex agglutination (LA)
- Virus neutralization (VN)

These assays work on the same basic principles: they rely on interactions between antibodies and antigens and require a known virus or viral protein, a patient sample (often serum), and an indicator. Antibodies will bind to the virus if present in the patient's sample.

Agar Gel Immunodiffusion (AGID)

In an agar gel immunodiffusion (AGID), also known as an agar gel precipitin (AGP) test, antigen–antibody complexes are formed by the diffusion of the virus and antibody through agar (a gelatin-like material), which forms a line of identity (Figure 3.1).

DOI: 10.1201/9781003369349-3

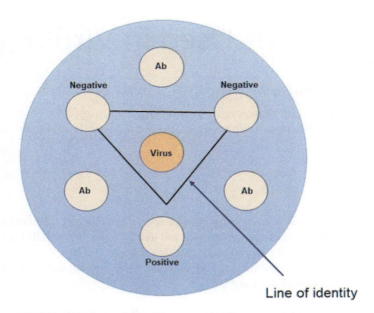

FIGURE 3.1 Schematic of an agar gel immunodiffusion (AGID) or agar gel precipitin (AGP) test. "Ab" represents a known antibody to the known virus in the middle.

Enzyme-Linked Immunosorbent Assay (ELISA)

The low cost and convenience of an enzyme-linked immunosorbent test, or ELISA, have made it a widely popular technology. Plastic wells are coated with either the target antigen (virus) or a protein specific to the target antigen (virus) of interest. After that, the sample (serum) is allowed to bind to the coated well, an antibody labeled with an enzyme is applied along with an indicator, and a color shift is observed. The presence of color indicates the presence of antibodies, and the absence of color indicates the absence of antibodies (Figure 3.2).

Hemagglutination Assay (HA) and Hemagglutination Inhibition (HI)

Many viruses have a surface protein that interacts with sialic acid receptors on the surface of red blood cells (RBCs) and can attach to them. The virus's surface protein is called hemagglutinin, and this property is known as hemagglutination. This creates a network, or lattice structure, of interconnected RBCs and virus particles. The concentrations of the virus and RBCs determine whether a lattice forms; if the relative virus concentration is too low, the lattice does not confine the RBCs, and they settle to the bottom of the well. Hemagglutination is observed in the presence of the influenza virus and several other viruses, as well as certain bacterial species such as staphylococci and vibrios. Based on the related surface receptors on the RBC and the selectivity of the targeted virus, the RBCs employed in HA and HI are usually from chickens, turkeys, horses, guinea pigs, or humans.

The HI test is based on the inhibition or blocking of hemagglutination. The influenza virus is one of the most well-known viruses that possess this characteristic. The patient's serum sample is incubated with the target virus, and red blood cells are added to the mixture of the virus and serum. The hemagglutination activity will be inhibited if antibodies are present; in the absence of antibodies, the virus will agglutinate, which means it binds together. The red blood cells serve as an indicator in this test.

Virus Neutralization (VN)

In the virus neutralization (VN) test, the sample of interest is incubated with the target virus, and changes in cell culture are observed, known as the cytopathic effect (CPE). Without antibodies, the virus will proliferate and cause CPE to be seen (Figure 3.3).

Methods Used in Virology

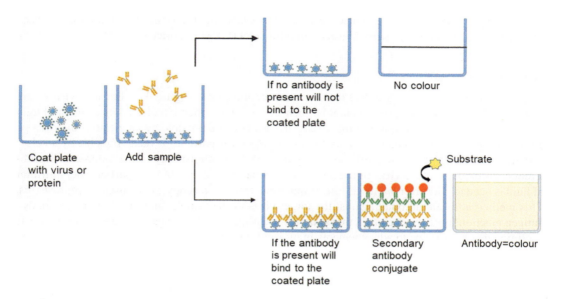

FIGURE 3.2 Enzyme-linked immunosorbent assay.

FIGURE 3.3 (Top) Normal baby hamster kidney cells-21. (Bottom) Baby hamster cells-21 infected with Newcastle disease virus showing cytopathic effect.

LIGHT AND ELECTRON MICROSCOPY

LIGHT MICROSCOPY

Unlike bacteria, viruses are too tiny to be seen under a typical light microscope. Consequently, antibodies intended to identify the target virus are utilized and labeled with an indicator, most commonly peroxidase or fluorescence. Because of this labeling, the virus cluster can then be seen using a light microscope (for peroxidase) or an ultraviolet (UV) light microscope (for fluorescence).

Electron Microscopy

The electron microscope is a useful tool for viral identification (Figure 3.4). Because viruses are much smaller than bacteria, they cannot be seen with the magnification offered by a standard light microscope. The ability to observe the virus particles is made possible by an electron microscope with 50,000× magnification. The ability to observe the virus particles is made possible by an electron microscope's 50,000× magnification (Figure 3.5). The insensitivity of this approach is problematic since, to detect the virus of interest, a concentration of about 10^6 viral particles per milliliter of fluid is needed. However, methods like immune electron microscopy can be used to improve this. This method adds an antibody against the target virus to the sample, and the resultant antibody–antigen reaction forms viral clumps that simplify visualization. The usage of electron microscopy is declining day by day due to advancements in molecular techniques.

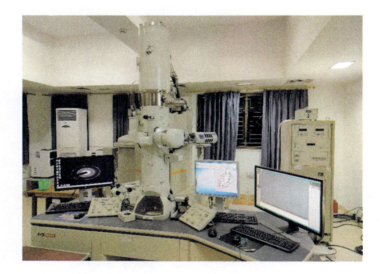

FIGURE 3.4 Field emission transmission electron microscope (FETEM).

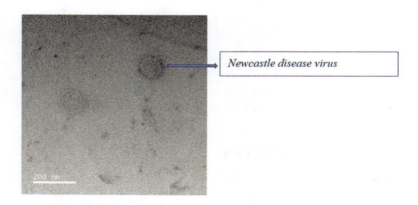

FIGURE 3.5 FETEM image depicting Newcastle disease virus.

Methods Used in Virology 15

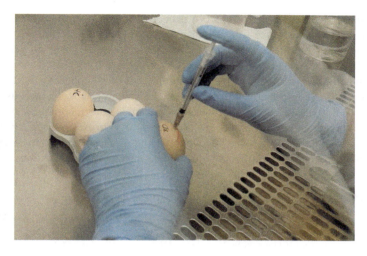

FIGURE 3.6 Inoculation in nine-day-old embryonated chicken eggs with Newcastle disease virus.

VIRUS ISOLATION AND MOLECULAR METHODS

Isolation of the virus is the first step in identifying a viral infection. Cell culture and fertile chicken eggs are the two most commonly used methods (Figure 3.6). Several problems are also associated with these methods, such as virus isolation being dependent on the viability of the virus particle. Since many viruses are unstable outside the host system, this method relies heavily on good sample handling and collection. Another issue is the availability of compatible systems, such as the appropriate cell line for propagating the suspected virus. It is necessary to have a range of options because not all viruses grow in the same environment or cell culture.

Most molecular techniques employed today are unique to the target of interest and are based on the development of an assay for a known virus target. However, isolation methods are less specific. Many viruses exhibit distinctive changes in the cells compared to the control cells as they grow in cell culture systems. Virus isolation is a labor-intensive and slow process. Hence, new tests are constantly being developed. Molecular techniques, such as polymerase chain reaction (PCR) and real-time PCR, are currently the most used assays (Figure 3.6).

MOLECULAR METHODS

Concerning the virology field, the area that is expanding the fastest is molecular methodologies and techniques. While virus isolation approaches require the presence of a live virus, molecular methods do not. These techniques increase the sensitivity of virus detection by identifying a portion of the viral genome.

The three conventional molecular methods are in situ hybridization, Southern blot, and dot blot. Specific DNA or RNA probes are used in these methods. Although they are more time-consuming and costly than some methods, they are highly sensitive.

The PCR and the reverse transcriptase PCR (RT-PCR), which identify DNA and RNA, respectively, and the advancement of more effective, standardized nucleic acid purification techniques have had the biggest impact on virology. The growing popularity of these techniques can be attributed to the quick developments and decreasing costs of nucleotide sequencing, commercial oligonucleotide manufacturing, and the accessibility of genetic sequences in public databases. The basic principle of PCR involves using two primers designed to identify a target sequence. If it exists, the target sequence of interest is amplified by a DNA polymerase and other elements needed for

DNA amplification. Since double-stranded DNA is needed for PCR, it can be used to identify DNA viruses. The discovery of the reverse-transcriptase enzyme capable of making a DNA copy of RNA enabled the application of PCR to include a reverse transcription step, which generates a double-stranded target (RNA–DNA), which is then used in PCR to identify RNA viruses.

There is still much to learn about virology's dynamic and ever-expanding field.

Beyond diagnosis, virological procedures are used in research labs. Numerous animal disease models are employed to study human diseases. Several molecular techniques, including cloning, inserting, and deleting genetic information, are used to modify viruses in such a way as to improve both human and animal health. Improved vaccinations and viruses designed to carry genetic material for gene therapy and cancer treatment are some outcomes of these alterations.

BIBLIOGRAPHY

Burrell, Christopher J., Colin R. Howard, and Frederick A. Murphy. *Fenner and White's Medical Virology*. Academic Press, 2016.

Castro, Anthony E., and Werner P. Heuschele. *Veterinary Diagnostic Virology: A Practitioner's Guide*. Mosby Year Book, 1992.

White, David O., and Frank J. Fenner. *Medical Virology*. Gulf Professional Publishing, 1994.

4 Transmission of Viruses

Shaurya Dumka and Sachin Kumar
Department of Biosciences and Bioengineering, Indian
Institute of Technology Guwahati, Guwahati, Assam, India

INTRODUCTION

Infection by viruses is quite different compared to other microorganisms. Before viruses can show their potential effect, they must enter the host's body. Viruses from various sources are released into the vicinity of the host cell environment, and the cells susceptible to these viruses allow the entry of a particular one. Susceptible cells are those that have a receptor for their complementary ligands. The virus must find new cells to infect if it is to survive. The different steps involved in the life cycle of a virus are (1) transmission and entry of viruses into the host system; (2) replication; (3) spread of the virus to adjacent and distant cells; and (4) shedding or persistence of the virus. This chapter will focus on the transmission of different viruses (vertebrates, invertebrates, and plants) and their entry routes (Carter and Saunders, 2007; Levinson, 2020).

Rivers, wind, and vectors can move viruses from one place to another. The classic example is the foot and mouth disease virus (FMDV) outbreak, which in 1981 moved over 250 km from its origin in Isle of Wight, UK, to Brittany via air. Viruses may also be transported from one host to another due to migration, travel, and animal export between different places. For example, the avian influenza virus due to bird migration, severe acute respiratory syndrome (SARS) due to human travel, and monkeypox virus due to animal export (Carter and Saunders, 2007).

Transmission can also occur between mother and offspring at the time of delivery (Zika virus, cytomegalovirus, hepatitis C virus [HCV], human immunodeficiency virus [HIV], human papillomavirus) or during breastfeeding (cytomegalovirus). Such transmission is vertical, whereas person-to-person transmission is horizontal (Levinson, 2020).

Viruses may also transmit from animals to humans and vice versa. The diseases caused by such types of transmission are known as zoonotic diseases. Animal-to-human transfer is known as anthropozoonosis, whereas human-to-animal transfer is known as zooanthroponosis or reverse zoonosis. This type of transmission could occur either directly from the bite of the reservoir host animal or indirectly through the bite of a vector (Dufossé, 2009). Some of the important zoonotic diseases caused by viruses are listed in Table 4.1.

VIRUS TRANSMISSION BY VECTORS: GENERAL PRINCIPLES

Organisms that carry the virus between different hosts act as vectors. Vectors may transmit the virus to plants, animals, and humans. Mostly, the vectors for viruses are of arthropod (insects, mites, ticks) origin, and sometimes the viruses transmitted by arthropods are also known as arboviruses (arthropod-borne viruses). The vector feeds on an infected host and transmits the virus to the uninfected new host. After ingesting the virions, they may cross the gut wall and ultimately reach the salivary gland of the vector. The saliva, now carrying the virus particles or virions, may transmit the virus to the healthy or new host while feeding, and this mode of transmission is known as circulative transmission. Vector-to-vector transmission may also occur sexually if the reproductive organs of the vectors are infected. Some viruses also transmit by transovarial transmission, which is sent to the next generation through the egg (Carter and Saunders, 2007).

DOI: 10.1201/9781003369349-4

TABLE 4.1

Zoonotic Diseases and Their Etiological Agents

Disease	Organism	Reservoir	Infective Hosts
Avian influenza	Influenza virus	Poultry, duck	Poultry and other bird
Hantavirus pulmonary syndrome (pneumonia)	Hantavirus	Rodents	Humans
Rabies	Rabies virus	Dogs, foxes, cats, bats	Humans and animals
Yellow fever	Yellow fever virus	Monkeys	Humans
Dengue	Dengue virus	Monkeys	Humans
Encephalitis	Equine encephalitis virus, West Nile virus, St. Louis encephalitis virus	Wild birds	Humans, horses, and other mammals

Source: Levinson (2020).

TABLE 4.2

Plant Virus Vectors

Plant Virus Vectors	Virus Transmitted
1. Insects	Cauliflower mosaic virus
• Aphids	Bean yellow mosaic virus
• Leafhoppers	Rice dwarf virus
• Whiteflies	Tomato yellow leaf curl virus
• Beetles	Maize chlorotic mottle virus
2. Mites	Ryegrass mosaic virus
3. Nematodes	Grapevine fanleaf virus

Source: Carter and Saunders (2007).

PLANT VIRUSES TRANSMISSION

Vectors play a significant role in the transmission of plant viruses since plant cells have a thick cell wall preventing the virus's direct entry. Most plant virus vectors are invertebrates (Table 4.2).

Aphids are the most common vectors of plant viruses. They ingest the contents of cells and transmit them during feeding on the plant cells. Most nematodes that transmit viruses to plants are soil-dwelling, and they enter the plant by piercing the root cells. An exciting feature about nematodes is that different shapes of viruses are transmitted by various types of nematodes specifically. For example, rod-shaped tobraviruses are transmitted by the Trichodoridae family of nematodes, while the Longidoridae family transmits isometric nepovirus virions. Many plant viruses bind specifically to the mouthparts of the favorable vectors via specific amino acid sequences present on the surface of capsids. Some parasitic fungi, for example, *Spongospora subterranean*, also act as plant virus vectors. *S. subterranean* causes powdery scab disease and is a vector of the potato mop-top virus. A separate genus *Furovirus* has been classified, reflecting the virion shape and transmission mode: *fungus* transmitted *rod*-shaped virus.

Plant viruses also show vertical transmission, i.e., the mother plant can infect the next-generation seeds. It mainly occurs in the embryo by an infected ovule or pollen grain. Nepoviruses and tobraviruses can be transmitted vertically. About 20 percent of plant viruses are transmitted vertically. Many plant viruses are also transmitted via artificial means, such as improper handling or grafting from an infected plant to a new host (Carter and Saunders, 2007).

Transmission of Viruses

VERTEBRATE VIRUSES

Nonvector Transmission

The most common route of infection in many vertebrates, including humans, is via the upper respiratory tract. The droplets shed by the infected individual during sneezing, coughing, and even speaking can infect a healthy individual. Apart from the respiratory route, the virus may be transmitted through the intestinal route, genital route, wounds, and many more. The fecal–oral route is the primary route of viruses infecting the intestinal tract. Ingesting contaminated food and water is responsible for intestinal viral infection. Some viruses, such as HIV and papillomavirus, are transmitted during sexual contact through genital secretions. Abrasions and wounds are other routes of transmission of viruses. Lesions present in the foot and mouth of an animal can lead to the transmission of FMDV from an infected to a healthy animal; similarly, herpes simplex virus (HSV) from lip lesion and papillomavirus from warts. These viruses are transmitted via direct contact between an infected and a healthy individual or animal, known as horizontal transmission. Rubella virus from mother to fetus via the placenta and HIV from mother to baby via milk are classic examples of vertical transmission of viruses among vertebrates. Table 4.3 lists different viruses transmitted by horizontal and vertical routes (Carter and Saunders, 2007).

Vector Transmission

Some vectors that transmit the virus to vertebrates are mosquitoes, mites, midges, and many other arthropods (Table 4.4). Out of these, the most common vector is the mosquito.

TABLE 4.3

Viruses Transmitted by Horizontal and Vertical Route

Route of Transmission	Virus Transmitted
1. Horizontal transmission	Influenza viruses
• Respiratory tract	Common cold viruses
• Intestinal tract	Measles virus
• Abrasions and wounds	Influenza viruses (birds)
• Genital tract	Rotaviruses
	Papillomaviruses
	Rabies virus
	HIV
	Papillomaviruses
2. Vertical transmission	Rubella virus
• Mother to fetus via the placenta	HIV
• Mother to baby via milk	

Source: Carter and Saunders (2007).

TABLE 4.4

Virus Transmitted by the Arthropod Vectors

Vector	Virus Transmitted
Mosquitoes	Yellow fever virus
	West Nile virus
Midges	Bluetongue virus
Ticks	Louping ill virus

Source: Carter and Saunders (2007).

ROUTES OF VIRUS ENTRY INTO THE HOST

The three main methods of transmitting viruses are the entry portal, replication within the host system, and an exit portal to infect a new healthy host. The viruses are mostly shed out from the same portal they entered. Some factors necessary for the establishment of the virus infection are:

1. Accessibility of the host cells to the virus.
2. Susceptibility of the host cells, meaning the cells must possess complementary receptors for the virus attachment. This affinity is sometimes known as tropism.
3. Permissibility of the cell toward an infection, meaning the cells must possess vital proteins and factors for viral replication to occur.

Susceptible and permissive cells must be in close contact with a virus to establish an infection. The host epithelial cells are the ones where most of the viruses interact. The epithelium is the main barrier between the external environment and the inner cavities of the body. The respiratory tract, gastrointestinal tract, and genital tract are all lined by a mucosal epithelium; the epithelium is covered by a protective layer of mucus (Louten, 2016; Tyler, 1999; Burrell et al., 2017) (Figure 4.1).

The four major transmission routes by which a pathogen can enter the human body are:

- Droplet and particle infection (influenza virus)
- Contact infection (HSV)
- Contaminated food and water infection (hepatitis A virus [HAV])
- Infection due to contaminated blood, tissue, or bodily secretions (HIV, direct transmission via blood; malaria, indirect transmission via insect bite) (Figure 4.2)

SKIN ENTRY PORTAL

The skin is the body's largest organ, creating 1.5–2 m² of surface area exposed to the external environment. The dense outer layer, keratin, forms the mechanical barrier against entering viruses. The low pH and fatty acid secretions on the skin provide further protection. Since the cells in the skin's outermost strata are dead, no viral replication can occur. To replicate, the virus must breach the skin via insect bites, cuts, punctures, wounds, and abrasions. Some viruses get confined to the skin while

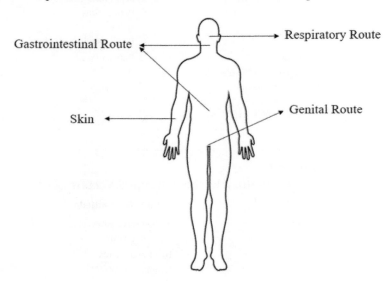

FIGURE 4.1 Different routes of entry in the human body.

Transmission of Viruses

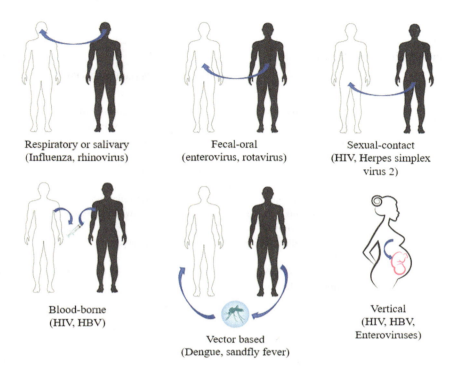

FIGURE 4.2 Various transmission modes (Burrell et al., 2017).

others can disseminate deeper. Viruses can reach the dermis and subcutis, finding the blood vessels, lymphatics, and nerve endings that can serve as routes of virus dissemination throughout the whole body. Insect bites are one of the most efficient ways of virus introduction through the skin, for example, equine infectious anemia (horses), myxoma virus (rabbits), and fowlpox virus (chickens). An animal bite can also transmit the virus from an infected to an uninfected host, for example, rabies. Viruses can also penetrate the skin via iatrogenic means, such as transmitted via hospital premises and instruments like needles, scissors, etc. A list of different viruses invading the skin is presented in Table 4.5 (Payne, 2017; Louten, 2016; Maclachlan and Dubovi, 2010).

Respiratory Tract Entry Portal

For many viruses, the respiratory tract is the primary entry portal. It serves as the means of entering and exiting gases from and into the environment. The respiratory system provides a large surface area where the viruses can enter and reside. The respiratory system is divided into the upper and lower respiratory tract. The viruses present in large droplets deposit in the upper part, whereas the particles in aerosols and liquids can pass into the lower respiratory tract. The epithelial cells on the surface of the respiratory tract are covered by a "mucociliary blanket," which is a layer of mucus (produced by goblet cells) and cilia on the luminal surface of epithelial cells lining the nasal mucosa and respiratory airways. These defenses play a significant role in minimizing the risk of infection by any pathogen. Larger particles (>10 μm in diameter) get trapped in this mucociliary blanket. In comparison, the smaller particles (<5 μm in diameter), like viruses, are inhaled directly into the lower respiratory tract and ultimately reach the lungs' alveoli, causing an infection. Most viruses get trapped in the mucus and, with the help of ciliary action, are carried deep into the respiratory airways and are either swallowed or coughed out. The respiratory system also consists of well-organized innate and adaptive immune mechanisms, including various lymphoid aggregates (NALT,

TABLE 4.5
Viruses and Their Entry Portal

Portal	Viruses That Use This Portal for Infection
Skin	
Direct contact	Human papillomaviruses, HSV-1, *Molluscum contagiosum virus*
Penetration into dermis or subcutaneous tissue	Injection/cuts: Hepatitis B virus, hepatitis C virus, HIV, Ebola virus
	Mosquitoes: Dengue virus, West Nile virus, eastern equine encephalitis virus, Chikungunya virus, yellow fever virus
	Ticks: Heartland virus, Powassan virus, Colorado tick fever virus
Respiratory tract	Adenovirus, measles, mumps, rubella, enterovirus D68, influenza A virus, influenza B virus, rhinovirus, respiratory syncytial virus, varicella zoster virus, variola, SARS
Gastrointestinal tract	Norwalk virus, rotavirus, poliovirus, enteric enteroviruses, hepatitis A virus, hepatitis E virus, sapovirus
Genital tract	Human papillomaviruses, HIV, hepatitis B virus, hepatitis C virus, herpes simplex virus-2 (HSV-2)
Eye	Adenoviruses, HSV-1, cytomegalovirus, enterovirus 70, Coxsackievirus A24, rubella virus, measles virus, vaccinia virus

Sources: Louten (2016), Mehraeen et al. (2021).

nasal-associated lymphoid aggregates; BALT, tonsils and bronchus-associated lymphoid tissue) that occur throughout the respiratory tract. Another kind of immune cell, alveolar macrophages, highly specialized in phagocytosis, is present in the lungs' alveoli. Despite having such protective mechanisms, viruses can attach to a particular receptor in the tract and infect the host, avoiding the clearance by the mucociliary blanket and immune-mediated responses. A list of different viruses invading the respiratory tract is presented in Table 4.5 (Leung, 2021; Klompas et al., 2021; Kutter et al., 2018; Mehraeen et al., 2021).

GASTROINTESTINAL ENTRY PORTAL

Many viruses come in contact with the stomach and intestinal epithelial lining by ingesting contaminated food and water. Relatively, the oral cavity and esophagus are less susceptible to viral infection. The gastrointestinal (GI) tract's first protective defense is the mucus lining the stomach and intestine mucosa, followed by acid production from the stomach, digestive enzymes' antimicrobial activity, and innate and adaptive immunity, such as immunoglobins (Igs). The B lymphocytes and mucosa-associated lymphoid tissue (MALT) produce the Igs. Despite such protective measures, viruses enter the GI tract via some specialized receptors or M (microfold) cells present in the Peyer's patches. The classic examples that cause enteric infections are rotavirus and enteroviruses. These are acid- and bile-resistant viruses. Many digestive proteolytic enzymes help viruses increase their infectivity by cleaving their capsid proteins and quickly entering the host cell (as in the case of rotavirus and some coronaviruses). The primary pathogenesis caused by such viruses is diarrhea, while some viruses are asymptomatic. A list of different viruses invading the gasrointestinal tract is presented in Table 4.5 (Maclachlan and Dubovi, 2010; Cliver, 1997; Wang et al., 2021).

SOME OTHER ROUTES

The genital tract is one route for some viruses to enter the body—for example, the herpes virus and HIV. Any abrasions in the penile mucosa or epithelial lining of the vagina may lead to virus transmission from an infected to a healthy individual during sexual intercourse. Conjunctiva can also be one route for transmitting some viruses like adenoviruses and enteroviruses (Maclachlan and Dubovi, 2010).

REFERENCES

Burrell, C. J., Howard, C. R., Murphy, F. A. J. F. & Virology, W. S. M. 2017. Epidemiology of viral infections. In *Fenner and White's Medical Virology* (pp. 185–203). Elsevier.

Carter, J. & Saunders, V. A. 2007. *Virology: Principles and Applications.* John Wiley & Sons.

Cliver, D. O. 1997. Virus transmission via food. *World Health Stat Q*, 50(1–2), 90–101.

Dufossé, L. 2009. *Encyclopedia of Microbiology.* Elsevier Amsterdam.

Klompas, M., Milton, D. K., Rhee, C., Baker, M. A. & Leekha, S. J. A. O. I. M. 2021. Current insights into respiratory virus transmission and potential implications for infection control programs: A narrative review. *Ann Intern Med*, 174(12), 1710–1718.

Kutterf, J. S., Spronken, M. I., Fraaij, P. L., Fouchier, R. A. & Herfst, S. J. C. O. I. V. 2018. Transmission routes of respiratory viruses among humans. *Curr Opin Virol*, 28, 142–151.

Leung, N. H. L. 2021. Transmissibility and transmission of respiratory viruses. *Nat Rev Microbiol,* 19, 528–545.

Levinson, W. E. 2020. *Review of Medical Microbiology and Immunology*, Sixteenth Edition, McGraw Hill LLC.

Louten, J. J. E. H. V. 2016. Virus structure and classification. In *Fenner and White's Medical Virology* (p. 19). Elsevier

Maclachlan, N. J. & Dubovi, E. J. 2010. *Fenner's Veterinary Virology.* Academic Press.

Mehraeen, E., Salehi, M. A., Behnezhad, F., Moghaddam, H. R. & Seyedalinaghi, S. J. I. D.-D. T. 2021. Transmission modes of COVID-19: A systematic review. *Infect Disord Drug Targets*, 21(6), 27–34.

Payne, S. J. V. 2017. *Introduction to Animal Viruses* (p. 1). Elsevier

Tyler, K. L. J. E. O. V. 1999. *Pathogenesis| Animal Viruses* (p. 1175). Elsevier.

Wang, C. C., Prather, K. A., Sznitman, J., Jimenez, J. L., Lakdawala, S. S., Tufekci, Z. & Marr, L. C. J. S. 2021. Airborne transmission of respiratory viruses. *Science*, 373(6558), eabd9149.

5 Virus Classification

Shinjini Bhattacharya, Sukanya Sonowal and Sachin Kumar
Department of Biosciences and Bioengineering,
Indian Institute of Technology Guwahati, Guwahati, Assam, India

INTRODUCTION

The systematic methods used to classify prokaryotes and eukaryotes are not very useful in virus classification, as viruses share no common ancestor. Virus classification was a subject of heated controversy around 1960, as their discoverers established no consistent system for naming viruses. If viruses represent relics of different organisms, genomic or protein analysis is not helpful because viruses have no gene sequence shared among them. For example, the 16S rRNA sequence, which is very useful in constructing prokaryotic phylogenies, is of no use in the case of viruses, as it carries no ribosome!

Biologists have used several classification systems in the past. In 1962, Robert Horne, Lwoff, and Paul Tournier put forward an extensive scheme for the classification of all viruses (bacterial, plant, and animal) under the classical Linnaean hierarchical system consisting of phylum, class, order, family, genus, and species. The primary principle that guided Horne and his colleagues in their classification system was that viruses should be grouped according to their shared properties rather than the properties of the cells or organisms they infect. A second principle focused on the type of nucleic acid genome a virus carries as the primary criterion for their classification. The importance of the nucleic acid was inferred from the Hershey–Chase experiment that proved that the viral genome alone can cause infection in the host. Four characteristics that formed the basis of virus classification in the past include:

- Symmetry of the protein shell (capsid)
- Nature of the nucleic acid in the virions (DNA or RNA)
- Presence or absence of a lipid membrane (envelope)
- Dimensions of the virion and the capsid

As per the report of the International Committee on Taxonomy of Viruses (ICTV) in 2005, approximately 40,000 virus isolates from bacteria, plants, and animals had been assigned to one of 3 orders, 73 families, 9 subfamilies, 287 genera, and 1,950 species. However, many viruses remain to be assigned or are assigned provisionally because they have not yet been characterized adequately.

VIRUS CLASSIFICATION BASED ON GENOME STRUCTURE AND CORE

Viral genomes vary in the type of genetic material they carry (DNA or RNA) and its organization (linear or circular, single- or double-stranded, and segmented or non-segmented) (Figure 5.1). Some viruses require additional proteins for their replication. Such proteins either remain associated directly with the genome or they are contained within the viral capsid. Examples of this classification are given in Table 5.1.

DOI: 10.1201/9781003369349-5

Virus Classification

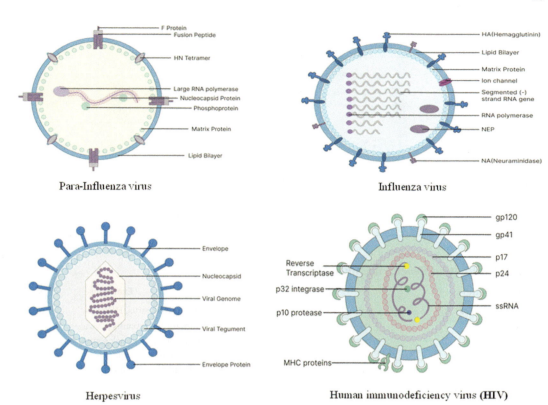

FIGURE 5.1 Diagrammatic representation of viruses based on genome structure and core.

TABLE 5.1
Virus Classification Based on Genome Structure and Core

Core Classification	Examples
DNA	Herpesvirus, smallpox virus
RNA	Retrovirus, rabies virus
Double-stranded	Herpesvirus, smallpox virus
Single-stranded	Rabies virus, retrovirus
Linear	Herpesvirus, rabies virus, smallpox virus
Circular	Retrovirus, rabies virus
Non-segmented	Parainfluenza viruses
Segmented	Influenza viruses

VIRUS CLASSIFICATION BASED ON CAPSID STRUCTURE

Viral capsids are a protein shell that surrounds the viral genome and protects it from digestion by enzymes. It also contains special sites on its surface that allow the virion to attach to host cells. Capsids are generally classified as naked icosahedral, enveloped icosahedral, naked helical, enveloped helical, and complex (Figure 5.2). Examples based on this classification are given in Table 5.2.

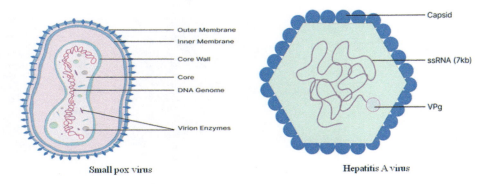

FIGURE 5.2 Diagrammatic representation of viruses based on capsid structure.

TABLE 5.2
Virus Classification Based on Capsid Structure

Capsid Structure	Examples
Enveloped icosahedral	Epstein-Barr virus, HIV-1
Naked icosahedral	Polioviruses, hepatitis A viruses
Enveloped helical	Influenza virus, measles virus
Naked helical	Tobacco mosaic virus
Complex with many proteins, some have a combination of both helical and icosahedral capsid structures	T4 bacteriophage, small pox virus

---

THE BALTIMORE SYSTEM OF VIRUS CLASSIFICATION

The most important function of a viral genome is to encode proteins that support its replication and assembly. However, the viral genome lacks the necessary protein synthesis machinery. Hence, an important principle is that all viral genomes must produce messenger RNAs (mRNAs) that can be translated by the host ribosomes (Figure 5.3). A second principle is that even with unfathomable time, evolution has led to the formation of only seven types of major viral genomes. The Baltimore system of classification, given by biologist David Baltimore in the early 1970s, integrates these principles to construct an elegant molecular algorithm for virologists all over the world. The refinement of the Baltimore system is that by knowing the basic nature of the viral genome, one can deduce the steps required to produce the mRNA. By the molecular biologist's convention, the strand containing the immediately translatable information is designated as a (+) strand. The DNA and RNA complements of (+) strands are designated as (−) strands.

> **Class I (Genome – dsDNA):** It is the most common type of animal virus. Viral DNA enters the nucleus of the cell, where the cellular enzymes transcribe the DNA and process the RNA into viral mRNA.
> **Class IA:** The DNA replicates inside the host cell nucleus, and the host enzymes support the DNA replication.

Examples: Adenoviruses, herpesviruses (varicella zoster, herpes simplex, and Epstein-Barr virus), papovaviruses (papilloma and polyoma).

> **Class IB:** The viral enzymes support the viral DNA replication in the host-cell cytoplasm. Example: Pox virus (smallpox, vaccinia, and cowpox).

Virus Classification

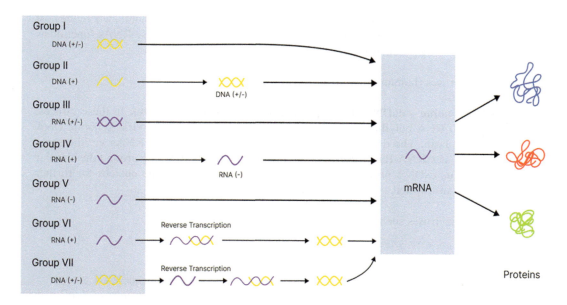

FIGURE 5.3 Diagrammatic representation of the Baltimore system of virus classification.

Class II (Genome – ssDNA): The ssDNA of these viruses is copied to dsDNA inside the host cell, which is then itself copied into mRNA.

Example: Parvoviruses. Some of these viruses encapsidate both positive and negative strands but in separate virion particles; others encapsidate only the negative strand.

Class III (Genome – dsRNA): The dsRNA genome is transcribed to mRNA by viral RNA-dependent RNA polymerase (RdRp). This mRNA is further used for replication and translation. The single-stranded mRNA is replicated to form the dsRNA genome.

Example: Reoviruses.

Class IV (Genome – plus sense (+) ssRNA): For (+) ssRNA viruses, the genome itself functions as mRNA. However, these viruses also produce positive sense copies of their genome from negative sense strands of an intermediate dsRNA genome. This acts as both a transcription and replication process since the replicated RNA is also mRNA.

Examples: Picornaviruses (polio virus and rhinovirus), togaviruses (rubella virus).

Class V (Genome – Minus sense (–) ssRNA): The genomic RNA of the virion in this case acts as a template for mRNA but does not encode proteins itself.

Examples: Rhabdoviruses (rabies), paramyxoviruses (measles and mumps), orthomyxoviruses (influenza).

Class VI (Genome – Plus sense (+) ssRNA that replicates with DNA intermediate): The linear genome of these viruses is converted to a dsDNA intermediate through a process called reverse transcription. The reverse transcriptase enzyme carried by the virus synthesizes a DNA strand from the ssRNA; this RNA strand is later degraded and replaced with a

DNA strand to create dsDNA. This dsDNA integrates into the host genome, where the host cell RNA polymerase II transcribes RNA in the nucleus. Some of these RNAs function as mRNA while others function as the genomic RNA of the virus.

Example: Retroviruses (human immunodeficiency virus [HIV]).

Class VII (Genome – dsDNA that replicates with RNA intermediate): In these viruses, the host cell RNA polymerase II transcribes RNA strands from the viral genomic DNA in the cytoplasm. The dsDNA is produced from these pregenomic RNAs via the same mechanism as ssRNA-RT viruses, but the replication occurs in a loop around the circular genome. This dsDNA produced can either be packed or sent to the nucleus for further rounds of transcription.

Example: Hepadnaviruses such as human hepatitis B virus.

BIBLIOGRAPHY

Maclachlan, N. James, and Edward J. Dubovi. *Fenner's Veterinary Virology.* Academic Press, 2010.
Racaniello, Vincent R., Glenn F. Rall, Anna Marie Skalka, and S. Jane Flint. *Principles of Virology.* John Wiley & Sons, 2015.

6 Virus Structure

Shinjini Bhattacharya, Sukanya Sonowal and Sachin Kumar
Department of Biosciences and Bioengineering,
Indian Institute of Technology Guwahati, Guwahati, Assam, India

INTRODUCTION

Viruses are minute and simpler in construction compared to other micro organisms; they contain only one type of nucleic acid, either DNA or RNA, but never both. Since viruses lack organelles such as ribosomes and mitochondria, they are completely dependent on their cellular hosts for protein synthesis and energy production. They are capable of replicating only inside the cells of their host. Virus particles are refined assemblies of viral and sometimes cellular macromolecules. These particles come in various sizes and shapes and vary largely in the nature of the molecules from which they are built. Virus particles are blueprints for the effective transmission of the nucleic acid genome from one cell to another within a single host or among host organisms.

VIRION

Progeny infectious virus particles that are formed by *de novo* self-assembly from the newly synthesized components within the host cells are called *virions*. The primary function of the virion is to protect the viral genomes, which are prone to be damaged irreversibly by a break or mutation in the nucleic acid during passage through hostile environments. During its transmission, a virus particle encounters many potentially dangerous physical and chemical agents that include nucleolytic and proteolytic enzymes, extremes of temperature and pH, and different forms of natural radiation. To survive such lethal exposures, the viral genome is sequestered within a hefty barrier formed by interactions among the viral proteins that form the protein coat. These interactions maintain very stable *viral capsids*, and virus particles comprising only protein and nucleic acid survive exposure to large variations in its environment. Some viruses possess an *envelope*, which is usually derived from the host cellular membranes and into which viral glycoproteins have been inserted. This envelope adds not only a protective lipid bilayer but also an external layer of protein and sugars that blocks the entry of enzymes or chemicals in an aqueous solution into virus particles.

The protective function of virus particles is solely dependent on the stable intermolecular interactions among their components during assembly, egress from the virus-producing cells, and transmission. However, these stable interactions must be reversed during virus entry and uncoating in a new host cell. The probable explanation of these paradoxical requirements suggests that the interaction of the virion with specific cell surface receptors and exposure to a specific intracellular environment can trigger substantial conformational changes. Thus, viruses can be described as metastable structures that have not yet achieved the minimum free energy conformation. Hence, virions are not inert structures but are molecular machines that play an active role in the delivery of the viral genome to the appropriate host cell and the initiation of the reproductive cycle.

Various functions of virion proteins are:

1. Protection of the viral genome
   - Assembly of a stable protective protein shell

DOI: 10.1201/9781003369349-6

- Specific packaging and recognition of the viral genome
- Interaction with the host cell membrane to form the envelope

2. Genome delivery
   - Specific binding to the host cell surface receptors
   - Transmission of specific signals that dictate uncoating of the genome
   - Inducing fusion with the host cell membrane
   - Interaction with the internal components of the cell is necessary to facilitate the transport of the viral genome to an appropriate location
3. Other interactions with the susceptible host
   - With cellular components established to establish a successful infectious cycle
   - With cellular components for facilitating transport to intracellular assembly sites
   - With the host immune system

NOMENCLATURE

Virus particles are complex macromolecular assemblies that are exquisitely suited for the protection and delivery of viral genomes. However, they are constructed according to the general principles of protein structure and biochemistry. The nomenclature used in explaining the virus structure is defined in Table 6.1.

FABRICATING A PROTECTIVE COAT

Nonetheless, despite their structural complexity, all virions contain at least one protein coat, either a nucleocapsid or a capsid, that protects and encloses the viral genome. In 1956, Francis Crick and James Watson suggested that most viruses appear to be rod-shaped or spherical under an electron microscope. Since viral genomes have limited coding capacity, it is proposed that the construction of capsids from a small number of subunits using helical symmetry or Platonic polyhedral symmetry would minimize the cost of encoding structural proteins. This criterion of genetic economy dictates that capsids and nucleocapsids be synthesized from identical copies of a small number of viral proteins possessing structural properties. The protein coats of all viruses, except for a few, display helical or icosahedral symmetry.

TABLE 6.1
Terminology Used in Description of Virus Structure

Term	Definition
Structural unit	Units forming capsids or nucleocapsids. It may include one protein subunit or multiple different subunits.
Subunit	A single, folded chain of polypeptides.
Morphological unit	Also called capsomeres, they are surface structures observed in electron microscopy.
Nucleocapsid	The nucleic acid–protein assembly packaged as a discrete substructure within the virion.
Capsid	The protective covering of protein shell encasing the viral genome.
Envelope	Lipid bilayer derived from the host cell carrying viral glycoproteins.
Virion	An infectious viral particle.

Virus Structure

CAPSIDS WITH HELICAL SYMMETRY

Helical symmetry is generally described by the number of structural units per turn of the helix. One characteristic feature of a helical structure is that any volume can be enclosed just by varying the length of the helix. The nucleocapsids of some enveloped animal viruses, certain plant viruses, and bacteriophages possess filamentous or rodlike structures with helical symmetry. The best-deciphered helical nucleocapsid, from a structural point of view, is that of tobacco mosaic virus (TMV), the first identified virus (Figure 6.1). TMV comprises a single molecule of (+) strand RNA, 6.4 kb in length, which is enclosed inside a helical protein coat. This protein coat is built from a single protein that folds into an elongated structure. Each coat protein molecule binds three nucleotides of the RNA genome in the interior of the helix. The coat protein molecules are involved in identical and equivalent interactions with the genome and with one another, hence allowing the construction of a large and stable structure from a single protein subunit.

The virus particles of several families of animal viruses with (–) strand RNA genomes, such as rhabdoviruses and paramyxoviruses, are reported to contain internal structures with helical symmetry enclosed within an envelope. Though these viruses have a common helical symmetry and similar composition, the internal components show considerable diversity in morphology and organization. Nucleocapsids of paramyxoviruses like the Sendai virus are filamentous, long structures in which the RNA packaging protein with the genome forms a left-handed helix containing a hollow core. In contrast, rhabdoviruses like the vesicular stomatitis virus have nucleocapsids that are squat, bullet-shaped structures closed at one end.

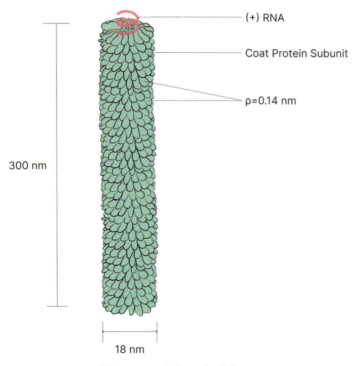

FIGURE 6.1 Representation of helical symmetry in tobacco mosaic virus.

CAPSIDS WITH ICOSAHEDRAL SYMMETRY

An icosahedron is defined as a solid having 20 triangular faces and 12 vertices related by two-, three-, and fivefold axes of rotational symmetry (Figure 6.2). In certain cases, virions are readily seen to be icosahedral, but most closed capsids and nucleocapsids appear spherical. These virions possess prominent surface structures, such as viral glycoproteins embedded in the envelope, that are not in conformation with the underlying icosahedral symmetry of the capsid shell. However, the symmetry with which the structural units interact is that of an icosahedron. The capsids with icosahedral symmetry are closed structures having a fixed internal volume.

Each face of the icosahedral capsid is formed by viral proteins. Viral proteins, however, are not triangular so a face is formed from at least three viral protein subunits fitted together. These can either be of the same protein or different proteins. The subunits jointly form the structural unit that repeats to form the capsid of the virion. Some viruses form very large icosahedral capsids, mostly due to the repetition of the structural unit. The number of structural units that makes each side is called the triangulation number (T); the structural units form the triangular face of the icosahedron. For example, in a T = 1 virus, one structural unit forms each icosahedron face. For a T = 4 virus, four structural units form the face. The proteins that comprise the structural unit may form three-dimensional structures known as capsomeres. These are clearly visible in electron micrographs. The capsomeres in icosahedral viruses take the form of pentons (containing five units) or hexons (containing six units), which are visible patterns observed on the surface of the icosahedron. Researchers have shown that proteins forming icosahedral symmetry require lesser amounts of energy; hence, these are evolutionarily more favored compared to other structures.

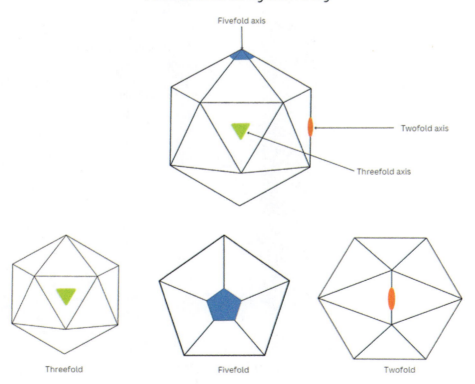

FIGURE 6.2 Representation of icosahedral symmetry.

Many animal viruses are icosahedral like human papillomavirus, herpesviruses, rhinoviruses, and hepatitis B virus. All icosahedral viruses and helical viruses can be naked or enveloped. However, the type of nucleic acid carried by the viruses does not correlate with their capsid structure.

COMPLEX VIRUS STRUCTURE

Though the majority of viruses can be classified on the basis of having either helical or icosahedral structures, a few viruses however have a complex architecture. These viruses do not conform strictly to the icosahedral shape or the helical shape. For example, geminiviruses, some bacteriophages, and poxviruses are viruses that possess a complex structure. Geminiviruses are plant-infecting viruses that are composed of two icosahedral heads joined together. Poxviruses are brick- or oval-shaped particles approximately 200–400 nm long. This complex virion contains a dumbbell-shaped core that encases the viral DNA, surrounded by two lateral bodies. The function of these lateral bodies is currently unknown. Bacteriophages are prokaryotic viruses that replicate within the bacteria. Many bacteriophages have complex structures; for example, bacteriophage P2 has an icosahedral head containing the genome attached to a cylindrical tail sheath for binding to the bacterial cell.

VIRUSES WITH ENVELOPES

Many viruses, in addition to capsids and nucleocapsids, contain an envelope formed by a viral protein-containing membrane derived from the host cell. These envelopes vary considerably in complexity, size, and morphology. The envelopes generally form the outermost layer of enveloped animal viruses. Moreover, the viral membranes also vary in their lipid composition, number of proteins, and their location. The viral envelope is embedded with proteins, mostly glycoproteins that carry covalently linked oligosaccharides. Carbohydrate moieties are attached to the proteins posttranslationally, during the transport to the cell membrane, the site at which the viral progeny assembles. The inter- or intrachain disulfide bonds that tend to stabilize the quaternary structure of the viral glycoproteins are also acquired during transportation to the assembly sites. These glycoproteins are mostly integral membrane proteins, having a short membrane-spanning domain embedded in the lipid bilayer. For example, the E protein of the flavivirus and the hemagglutinin (HA) protein of the human influenza A virus differ structurally but both are primed for a dramatic conformational change allowing the entry of internal components of the virion into a host cell.

PACKAGING OF THE VIRAL GENOME

The presence of nucleic acid is one of the definitive properties of the virion. The incorporation of nucleic acid inside the virus particle during its assembly requires discrimination from a large population of cellular nucleic acid. Since the volume of the viral capsid/nucleocapsid is finite, condensation of the viral genome is necessary for accommodation. However, packaging is an intrinsically unfavorable process as the highly constrained conformation of the nucleic acid results in a loss of entropy. In certain cases, specialized viral proteins drive the insertion of the genome by harnessing energy from ATP hydrolysis. Three general mechanisms for packaging the viral genome are:

- By direct contact with capsid or nucleocapsid proteins
  Examples: Herpesvirus, parvovirus, reovirus
- By cellular DNA-binding proteins
  Examples: Rhabdovirus, polyomavirus, retrovirus
- By specialized viral nucleic acid proteins
  Examples: Poxvirus, adenovirus, orthomyxovirus

METHODS USED TO STUDY VIRUS STRUCTURE

The most common methods used to study virus morphology and structure are electron microscopy. This technique depends on the negative staining of the purified virus particles with an electron-dense substance such as phosphotungstate or uranyl acetate. This provided the first rational basis of virus classification. The minimum size of an object that can be distinguished by classical electron microscopy is limited to 50 to 75 Å even when the virus structure is well preserved. However, cryo-electron microscopy can provide improved resolution to 10 to 20 Å.

BIBLIOGRAPHY

Louten, Jennifer. "Chapter 2 - Virus Structure and Classification." In *Essential Human Virology*, edited by Jennifer Louten, 19–29. Boston: Academic Press, 2016.

Maclachlan, N. James, and Edward J. Dubovi. *Fenner's Veterinary Virology*. Academic Press, 2010.

Racaniello, Vincent R., Glenn F. Rall, Anna Marie Skalka, and S. Jane Flint. *Principles of Virology*. John Wiley & Sons, 2015.

7 Virus Replication

Shinjini Bhattacharya, Sukanya Sonowal and Sachin Kumar
Department of Biosciences and Bioengineering, Indian
Institute of Technology Guwahati, Guwahati, Assam, India

INTRODUCTION

A fundamental principle separating viruses from other replicating entities is the fashion in which new virus particles are synthesized. Virus particles are assembled *de novo* from different structural components that are independently but somewhat synthesized in synchronized events. This unique replication pattern was recognized earlier from studies using bacteriophages. The experimental outline is as follows:

- Add a chloroform-resistant phage to a culture of bacteria for several minutes.
- Rinse well to remove all non-attached phages, followed by incubation.
- Remove samples at various time points.
- Add chloroform to the samples to stop growth.
- Quantify the number of phages at each time point.

This experiment will give the "one-step growth curve," which can be achieved with any type of virus (Figure 7.1). The most important finding of this type of study is that the virus disappears from the infectious cultures for a variable period of time. This is referred to as the eclipse period, which represents the time required for different components of the virion to be synthesized and assembled. After the beginning of the assembly, there is an exponential increase in the infectious virus particles. This continues until the host cell is unable to maintain its metabolic integrity. The virus may either be released suddenly by lysis of the host cell (T-even bacteriophage), or it may be released slowly when the maturation of the virus occurs at a cell membrane site (influenza virus).

The basic components of a virus replication cycle are:

1) Attachment
2) The eclipse phase, which includes penetration, uncoating, replication and synthesis of viral components, and maturation
3) Release of virus particles

A generalized overview of viral replication is illustrated in Figure 7.2.

ATTACHMENT

The first step critical to the virus replication cycle is the binding of the virus with the host cell. The series of interactions involved in the binding process defines the host range of the virus and tissue or organ specificity, which is tropism. The tropism largely defines the viral pathogenic potential and the nature of the disease it induces. Viruses have evolved to use the host cell surface molecules that are otherwise critical for different cellular processes. The initial interaction of the virion with the cell surface might involve short-distance electrostatic interactions with charged molecules such as heparan sulfate proteoglycans. This initial contact may be a simple strategy to concentrate the virus

DOI: 10.1201/9781003369349-7

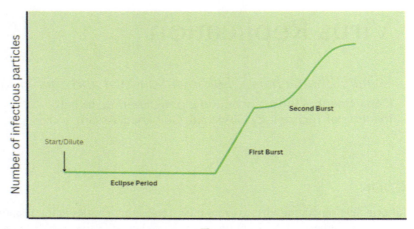

FIGURE 7.1 One-step growth curve of viruses.

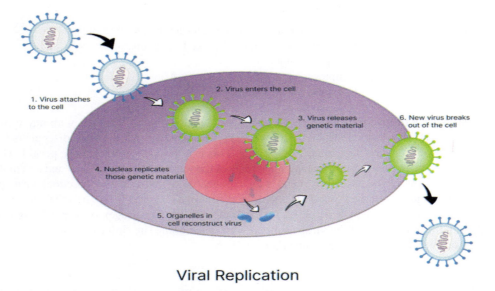

FIGURE 7.2 An overview of viral replication.

on the cell surface, allowing more specific interactions with other receptor molecules. The attachment is a temperature-independent process compared to penetration, which depends on the fluidity of the lipid membrane and has temperature constraints. A few examples of cellular macromolecules used by viruses as entry receptors are summarized in Table 7.1.

Many viruses require more than one entry factor to establish a successful infection within the host cell, making the virus–host receptor interaction more complicated. A classic example of this phenomenon is the human immunodeficiency virus (HIV), where the initial interaction occurs through heparan sulfate followed by binding to the CD4 receptor and a chemokine receptor such as CCR5 or CXCR4. The highly restricted host range of some viruses is due to the requirement of a unique entry factor found only on highly differentiated cell types.

Virus Replication

TABLE 7.1

Receptors Used by Viruses for Entry into the Host Cell

Virus	Receptor
Poliovirus	PVR (CD155)–Ig family
Coxsackievirus B	CAR (coxsackie and adenovirus receptor)–Ig family
Avian leucosis/sarcoma virus	Tissue necrosis factor-related protein-TVB
Human rhinovirus 14	ICAM-1 (intercellular cell adhesion molecule 1)–Ig family
Rotavirus	Various integrins
Influenza A virus	Sialic acid
Herpes simplex virus 1	HveA (herpes virus entry mediator A), heparan sulfate proteoglycan
Adenovirus 2	CAR–Ig family
Epstein-Barr virus	CD21, complement receptor 2 (CR2)
Foot-and-mouth disease virus (wild type)	Various integrins

VIRAL ENTRY INTO THE HOST CELL (PENETRATION)

Following the viral attachment, viruses must enter the host cells. This requires the virus to traverse the lipid bilayer surrounding the host cell. However, the detailed mechanisms of penetration for many viruses are still unclear. The three modes of entry employed by most viruses are explained next.

DIRECT PENETRATION OF THE VIRAL GENOME

In this method, the host cell cytoplasmic membrane and the viral capsid cause a rearrangement of capsid proteins, allowing the nucleic acid of the virus to pass through the membrane into the cytoplasm. For example, the poliovirus, a non-enveloped virus, upon binding to its receptor on the host cell surface, forms a pore in the host cell membrane to inject its RNA genome.

DIRECT FUSION WITH THE PLASMA MEMBRANE

This method is commonly used by enveloped viruses. The viral envelope fuses with the host cell plasma membrane, triggering the release of the nucleocapsid into the host cell cytoplasm. For example, HIV fuses directly with the host cell cytoplasmic membrane.

RECEPTOR-MEDIATED ENDOCYTOSIS

This method of penetration is used by both naked and enveloped viruses. In this process, the host cell membrane invaginates and pinches off, placing the virus in an endocytic vesicle. Enveloped viruses, such as the influenza virus, first bind to the host cell surface receptors and induce receptor-mediated endocytosis. The endosome acidifies further, resulting in the fusion of the viral envelope with the endosomal membrane. This results in the release of the nucleocapsid into the cytosol. Adenovirus, another non-enveloped virus, enters the cell by inducing receptor-mediated endocytosis and disrupts the endosomal membrane to release the capsid into the cytosol.

The entry of the nucleocapsid or the viral genome into the host cell cytoplasm does not reflect the final step of the replication initiation process for the virus. In certain cases, the genome is not in a form to serve as a template for the synthesis of mRNA, which is required to direct the production of viral proteins. Many cellular processes are further involved in the translocation of the viral unit

TABLE 7.2

Characteristics of Viral Propagation in Different Families

Family	Site of Nucleic Acid Replication	Site of Maturation	Uptake Route
Poxviridae	Cytoplasm	Cytoplasm	Variable
Herpesviridae	Nucleus	Nuclear membrane	Variable
Adenoviridae	Nucleus	Nucleus	Clathrin-mediated endocytosis
Asfaviridae	Cytoplasm	Plasma membrane	Clathrin-mediated endocytosis
Polyomaviridae	Nucleus	Nucleus	Caveolar endocytosis
Papillomaviridae	Nucleus	Nucleus	Clathrin/caveolar endocytosis
Retroviridae	Nucleus	Plasma membrane	Plasma membrane fusion or clathrin-mediated endocytosis
Rhabdoviridae	Cytoplasm	Plasma membrane	Plasma membrane fusion
Paramyxoviridae	Cytoplasm	Plasma membrane	Plasma membrane fusion
Hepadnaviridae	Nucleus/cytoplasm	Endoplasmic reticulum	Clathrin-mediated endocytosis

in the proper orientation to the correct location for its replication. This translocation process might involve the microtubule transport system to enable more localized movement.

REPLICATION OF THE VIRAL GENOME

Up until now, the virus has been in a passive state, for the most part, and no biosynthetic activity directed by the viral genome has occurred yet. The preliminary steps have put the viral genome in a proper position so that it can initiate a successful replication cycle. This reprograms the cell to assist in the production of mature virus particles. The next phases of the replication cycle are unique to each virus family; examples of a few are outlined in Table 7.2.

REPLICATION OF DNA VIRUSES

After penetration, the viral nucleic acid is prepared for expression and replication. In DNA viruses, early viral proteins are involved in hijacking the host cell machinery and the synthesis of viral RNA and DNA. DNA replication usually occurs in the host cell nucleus with exceptions such as poxviruses that replicate solely in the cytoplasm. An overview of DNA virus replication is illustrated in Figure 7.3.

Unlike other DNA viruses, hepadnaviruses (hepatitis B virus) are unique as they contain a partially dsDNA genome that must be converted into RNA form by the reverse transcriptase enzyme carried by the virion during the virus's life cycle. After infection, the viral DNA is released into the nucleus. The transcription of viral genes occurs inside the nucleus using host cell RNA polymerase and yields mRNAs. The RNA moves out into the cytoplasm, where it is reverse transcribed into minus-strand DNA. The RNA strand is degraded by RNase H, and the remaining fragment of the RNA serves as a primer for the synthesis of the gapped dsDNA genome, using the minus-strand DNA as a template.

REPLICATION OF RNA VIRUSES

An overview of RNA virus replication is illustrated in Figure 7.4. RNA genome-containing viruses are much more diverse in their replication strategies compared to DNA viruses. Most RNA viruses

Virus Replication

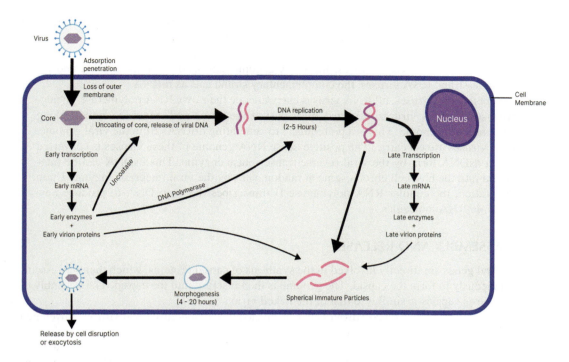

FIGURE 7.3 An overview of poxvirus replication.

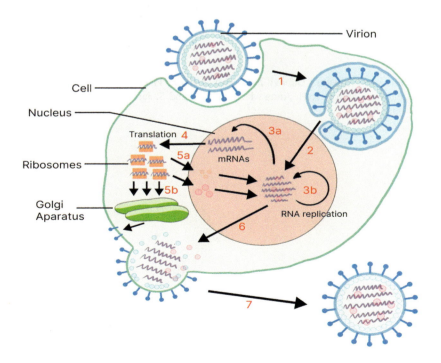

FIGURE 7.4 An overview of influenza virus replication. (1) Binding. (2) Entry. (3a and 3b) Complex formation and transcription. (4) Translation. (5) Secretion. (6) Assembly. (7) Release.

can be placed in one of four general groups depending on their modes of genome replication and transcription, and also based on their relationship to the host genome.

- For positive-strand RNA viruses, the RNA genome can itself serve as mRNA, and for negative-strand RNA viruses, the complementary strand acts as mRNA.
- Except for retroviruses, all RNA viruses carry/encode their own RNA polymerase. In negative-strand RNA viruses, this enzyme is part of the virion and enters the cytosol along with the viral genome. This is a prerequisite to generating mRNAs from the viral genome.
- Retroviruses are eccentric. The positive-sense RNA genome of these viruses is converted into a dsDNA form by the viral reverse transcriptase enzyme. This dsDNA can be integrated into the host cell chromosome at random sites by the viral integrase enzyme. Upon integration, the cellular RNA polymerase II transcribes the viral DNA to produce new genomic RNA molecules.

VIRAL ASSEMBLY AND RELEASE

Specific viral genes are directly involved in the synthesis of capsid proteins, which then self-assemble spontaneously to form the capsid. The genome is then inserted into the capsid. The assembly of enveloped virus capsids is similar to that of the naked virion.

The mechanisms of viral release differ between enveloped and non-enveloped viruses. Enveloped viruses are released from the host cell by budding. Most of the viral envelopes arise from the host cell membranes by a multistep process. Most envelopes arise from the plasma membrane, except in herpes viruses where the envelope formation also involves the nuclear membrane. The viral proteins and glycoproteins are incorporated into the host cell membrane prior to budding. During budding, the host cell membrane embedded with viral glycoproteins and proteins pinches off to form the viral envelope.

BIBLIOGRAPHY

Maclachlan, N James, and Edward J Dubovi. *Fenner's Veterinary Virology.* Academic Press, 2010.
Racaniello, Vincent R., Glenn F. Rall, Anna Marie Skalka, and S. Jane Flint. *Principles of Virology.* John Wiley & Sons, 2015.

8 Viral Pathogenesis

Aditya Singh Chauhan, Archita Sen and Sachin Kumar
Department of Biosciences and Bioengineering, Indian
Institute of Technology Guwahati, Guwahati, Assam, India

VIRAL PATHOGENESIS

Viral pathogenesis is the mechanism by which a virus causes disease in a host. The severity and form of the disease caused by the virus are determined by interactions both at the systemic and cellular levels between the host and the virus. Viruses range from giant DNA viruses like herpes and poxviruses to small RNA viruses like flaviviruses (*Dengue virus*) that showcase distinctive interactions with the host to establish disease. Virus-specific differences in outcomes result from factors like tissue tropism, modes of transmission, interactions with host immunity, and differences in viral replication cycles, as well as several other variables such as differences in host immune status, host genetic variation, route of inoculation, viral dose, and differential individual reaction. Even though each virus has its unique pathogenesis and pathology, several of the viral life cycle stages/disease processes are common among all the viral pathogens (Baron et al., 1996).

IMPLANTATION OF THE VIRUS

Viruses can spread through various channels, including bites, contaminated objects, food, and air. As a result, they could implant on any surface of the body and internal sites. Viruses have the highest chances of implantation at sites where there is direct contact between the cells and the virus, such as the alimentary tract, the respiratory tract, subcutaneously, and the genital tract. Virus implantation can also happen in the fetus during fertilization through infected germ cells, during gestation through the placenta, or even at birth. Several factors, including the dosage, infectivity, and virulence of the virus implanted and the implantation site, can influence the severity of an infection and its potential to manifest as a moderate, severe, or fatal condition, even in the early stages of pathogenesis/disease (implantation) (Baron et al., 1996).

REPLICATION OF THE VIRUS LOCALLY AND ITS SPREAD

Local viral dissemination and replication may occur after the successful implantation of the virus. The spread of the virus to adjacent cells may occur both extracellularly and intracellularly after replication inside the initially infected cells. A virus can propagate extracellularly by infecting a neighboring cell after it is released into the extracellular fluid. Within cells, the virus spreads by cytoplasmic bridges or by merging infected cells with nearby uninfected ones.

Although most viruses transmit extracellularly, some viruses, like poxvirus, paramyxovirus, and herpesvirus, can transmit both extracellularly and intracellularly. Intracellular spread is more favorable and gives viruses a more protective environment as the neutralizing antibodies against the virus are unable to pass through the cell membrane. Extracellular spread to cells that are immediately outside and in the vicinity of the initially infected cell can happen by diffusion of surface fluids like the mucous layer of the respiratory tract or by liquid spaces present at the local site, such as lymphatics. Moreover, viruses can also spread within the local tissue via migratory cells like infected lymphocytes and macrophages. After the infection has been established at the virus's entry

DOI: 10.1201/9781003369349-8

point, it is followed by local viral multiplication that results in localized illness and virus shedding. Thus, in a number of infections, the local implantation sites may serve as both the target organ and the site of viral shedding. Examples of such infections are certain respiratory tract infections like influenza, the common cold, and parainfluenza infections, and certain alimentary tract infections caused by picornavirus and rotavirus. Localized skin infections that fall into this category include cowpox, warts, and molluscum contagiosum. Some infections also have the potential to spread to distant body surfaces from their localized infection site. The conjunctivitis epidemic caused by the picornavirus illustrates this; the virus can also propagate in the absence of viremia (detectable viral presence in the blood) straight from the eye to the pharynx and intestine. Other viruses spread internally to far-off target organs and excretory sites. There is also a third class of viruses that are capable of producing both local and disseminated viral infections like measles and herpes simplex (Baron et al., 1996).

VIRAL SPREAD IN THE BLOODSTREAM

Blood and peripheral nerves that are in contact with the reproducing virus at the site of entry serve as the primary routes for widespread dissemination of the virus throughout the body. The bloodstream is the most typical method of viral systemic propagation. After an initial replication period at the portal of entry, viruses such as those that cause measles, poliomyelitis, and smallpox are spread through the blood. However, these viruses cause no significant symptoms or signs of illness at the site of entry, as the cells that are killed at this site are expendable and easily replaced. After the initial replication, the progeny of the virus spread to the lymphoid tissue via the afferent lymphatics and then to the endothelial cells, particularly those of the lymphoreticular organs, in close proximity to the bloodstream, via the efferent lymphatics. A transient primary viremia could arise from this early spread. Following this, a secondary viremia occurs when the released virus enters the bloodstream directly and exposes all body tissues and capillary systems, typically lasting several days. The target organs can be infiltrated by a virus from the capillaries after its replication inside the capillaries' fixed macrophages and endothelial cells, after which it gets deposited/released on the target organ side of the capillary. Viruses may also permeate through tiny openings in the capillary endothelium or utilize the migrating leukocytes that are infected to penetrate the capillary wall. After the virus has reached the target organ, it reproduces and spreads within the target organ or location of excretion using the processes that cause local dispersion at the portal of entry. The diseased state occurs only if the virus replicates in a significant enough number of vital cells and destroys or damages them. For instance, in poliomyelitis, the alimentary canal serves as both the entrance point and the site of shedding, while the central nervous system is the target organ where the virus causes the infection. The target organ and the place of shedding may also coincide in some circumstances (Baron et al., 1996).

VIRAL SPREAD IN THE NERVES

Nerve dissemination contributes to the spread of some severe disorders, albeit being less common than viral dissemination through the bloodstream. The poliomyelitis, herpes, and rabies viruses cause illnesses along this pathway. For example, the rabies virus that infects humans after an animal bite can multiply within the muscle tissues and subcutaneous spaces before accessing the nerve endings. The rabies virus appears to spread centrally in neurites (axons and dendrites) and perineural cells, where it is protected from antibodies. The rabies virus uses this neural pathway to enter the central nervous system, the site of origin of the disease. After that, the rabies virus travels centrifugally across the neurons and sheds in the salivary glands (Baron et al., 1996).

Viral Pathogenesis

INCUBATION PERIOD

Viral infections typically do not show any symptoms or signs of disease until the propagation stage of the virus. The interval between being exposed to a virus and the development of illness is known as the incubation period, and it begins at the moment of implantation and continues during the dissemination phase until the virus replicates in the target organs and causes disease. When the virus is present in the bloodstream, malaise and a minor fever can occasionally occur, although they are mostly transient and do not have any major diagnostic significance. The length of the incubation period depends on how far a virus has to travel to infect the target organ; for short distances, the incubation period is short, lasting up to one to three days. However, in widespread infections where the virus has to travel sequentially through the body to reach the target organs, the incubation periods are longer. Other variables may also influence the incubation period. Viruses that are injected intravascularly via insect bite, like the togavirus causing generalized infections, are fast proliferating and tend to exhibit a very brief incubation period. It is generally unclear what processes control the prolonged incubation time of persistent infections, which can last from months to years. In prolonged infections, the lysis of the persistently infected cell frequently occurs slowly or not at all (Baron et al., 1996).

TISSUE TROPISM

As obligatory intracellular parasites, viruses cannot proliferate outside their host cells. Most viruses are limited to particular cell types in particular organs, rather than infecting the entire host. Tropism is the range of tissues that a virus can infect. For instance, a neurotropic virus replicates in brain cells, but an enterotropic virus does so in the stomach. Some viruses are pantropic, meaning they can replicate and infect a wide range of tissues and cell types.

Therefore, a virus's ability to cause disease and the kind of disease it causes are primarily determined by the type of cells it targets and the extent to which the targeted cells are impacted by the viral infection. Certain viruses do not directly destroy the infected cell, whereas other viruses are cytopathic, meaning that the virus infection causes the host cell to die. Certain viruses target and kill vital cell types, such as neurons, to cause disease. A minimum of four factors determine tropism. Firstly, it can be ascertained by the distribution of entry receptors (susceptibility), and, secondly, by the virus's need for differently produced intracellular gene products to finish the infection (permissivity). Even in cases where the cell is susceptible and permissive, the viral infection can be prevented if the virus is physically unable to come in contact with the tissue (accessibility). Lastly, viral infection can still fail in spite of the cell being susceptible, accessible, and permissive due to the presence of the host's local innate immune responses. Tropism is typically defined by combining two or more of these factors (Baron et al., 1996).

RECEPTOR INTERACTIONS

Viruses enter host cells by attaching to specific cell surface receptors through surface molecules. Sometimes, a single molecule on the cell surface interacts with a virus to cause both viral binding to the cell and viral entry. For instance, ICAM-1 (CD54) receptors are used by several rhinoviruses for binding, and such interactions encourage infection of the cell by the virus (Albrecht et al., 1990). On the other hand, some viruses interact with cell surface chemicals to attach to the host cell, like the herpes simplex virus, which interacts with heparin sulfate on the cell surface, followed by binding to specific host receptor proteins that facilitate viral entry into the cell (Spear, 2004). Such interactions between the virus and receptors can have a significant influence on a number of features of viral pathogenesis, such as the extent to which the virus infects and causes disease in a specific host, and also has an impact on the tissue and cell tropism. The intricate interplay between viruses and the receptors on host cells highlights the critical role these interactions play in disease pathogenesis.

In addition to impacting disease development and viral cell tropism, interactions between a virus and its receptor are crucial in determining a virus's ability to infect new hosts, particularly in the case of zoonotic viruses. Although these viruses usually live in a particular host species, they can spread to other species and cause illness. To successfully transition from their natural host to a new one, zoonotic viruses must either engage with highly conserved, species-specific receptors or undergo genetic modifications that enable them to bind to receptors in their new host. An example of a zoonotic virus is the Sindbis virus, transmitted by mosquitoes, which utilizes a highly conserved receptor protein across species, known as the natural resistance-associated macrophage protein (NRAMP), to cause infection in both mosquitoes and vertebrate hosts (Rose et al., 2011). In contrast to virus interaction with highly conserved receptors, some viruses must interact with different receptors in different species. Therefore, for the virus to successfully traverse species barriers, it will probably need to adapt so that it can better exploit the receptor in the new species. This type of interaction is exemplified by the SARS coronavirus, or SARS-CoV, which causes severe acute respiratory illness by using the interaction between the spike protein (S) of the virus and the angiotensin-converting enzyme-2 (ACE2) of the cells (Li et al., 2003). Nevertheless, it is believed that SARS-CoV typically lives in bats, and the virus generated from these animals does not effectively interact with human ACE2. However, certain alterations in its receptor binding region resulted in more effective interactions with the human ACE2 protein. These adaptive mutations made the viruses more capable of infecting human cells. This is corroborated by the discovery that the introduction of a portion of the human ACE2 binding spike protein into the bat SARS-CoV virus can increase its infectivity to humans. These findings support the hypothesis that interactions between viruses and host cell receptors are critical in determining the ability of viruses to infect and cause disease in a host species (Li et al., 2003).

VIRAL PERSISTENCE AND LATENCY

STATES OF VIRUS INFECTIONS

Viral infection involves the insertion of the virus's genome into the host cell. This interaction could have several outcomes, depending on the virus's ability to replicate inside the cell and produce infectious progeny. When a new infectious virus is created, an infection is considered productive; if none, it is considered abortive. If an infectious virus is not produced immediately but still can cause a productive infection, the infection is considered latent. Reactivation is bringing an infectious cycle back to life from a latent condition. Latency, on the other hand, is not merely a sluggish, productive replication cycle, but it also denotes a distinct transcriptional and translational state in which the infectious virus is absent but in which, if it becomes necessary, a productive replication cycle can be restarted. If a cell can support a productive infection, it is permissive; if not, it is nonpermissive (Baron et al., 1996).

ACUTE AND PERSISTENT INFECTIONS

Acute infection occurs in the initial stage of the viral infection when the virus gains entry into the susceptible host. In contrast, persistent, or chronic, infection takes place when the host's immune system is inefficient in clearing the viral infection in a time frame typically expected for the clearance of acute infections, therefore leading to prolonged infection. Several mechanisms, including reactivation, latency, continued replication, and nucleic acid persistence, cause chronic infection. Many of these processes can happen simultaneously. Long-term detection of viral nucleic acid in the host has been observed in certain instances, albeit the nature of the infectious process is yet unknown. In these situations, chronic infection could refer to a latent infection, an abortive infection without the removal of leftover nucleic acid, continuing replication, or possibly an undiscovered viral infection. In certain instances, such as with hepatitis B virus (HBV) or hepatitis C virus

Viral Pathogenesis

(HCV), some people develop a persistent infection while others recover. In these situations, the point at which most patients no longer have acute infections is often characterized as the shift from acute to chronic. In some situations, such as with herpesviruses or lentiviruses like the human immuno-deficiency virus (HIV), almost all hosts develop a persistent infection. In this instance, the period needed to clear the initial wave of viral replication and create equilibrium between the virus and the host defines the change from acute to chronic infection. Continuous replication and the creation of latency are the two main mechanisms leading to chronic infection development. The virus uses a genomic and transcriptional approach, frequently requiring restricted viral gene expression during latent viral infection to enable the genome to survive without lytic replication. Examples include herpesviruses like Epstein-Barr virus (EBV) and herpes simplex virus (HSV), which exhibit selec-tive expression of their viral genes in their proviral or circular episomal form. Latent infection is often immunologically silent because infected cells do not express any viral proteins. Given that the host lacks any known means of detecting the virus's existence, this is the most extreme type of immune evasion. The virus must be able to reactivate and restart the lytic cycle of gene expression to live and propagate from the latently infected cell. This could result in the production of antigens that the immune system can recognize and react to. A virus must continue producing infectious viri-ons despite innate and adaptive immune responses to thrive through continual replication. It may be necessary for viruses to periodically reactivate and replicate to maintain latency and propagate from host to host, even for those that use the maintenance of latency as their primary tactic. Certain viruses, like HIV, survive by establishing latency and continuing to replicate continuously, making them very challenging for the host's immune system to combat (Baron et al., 1996).

HOST FACTORS INFLUENCING VIRAL PATHOGENESIS

PROVIRAL FACTORS

Factors other than receptor interactions can play a role in determining the tissues a virus might infect. Even though the majority of viruses are able to encode their own replication machinery, they nevertheless rely on the host cell for several activities, such as those that facilitate viral entrance or assembly and translation of viral proteins. Host cells also harbor certain molecules that facilitate viral infection; such molecules are called proviral host factors. Numerous proviral factors are part of generally significant cellular processes, like cellular protein transport pathways or the host trans-lation machinery, which are probably crucial for large classes of viral diseases. Nonetheless, there are situations in which host variables combine with particular viruses to promote viral replication in particular cell types, which influences the pathophysiology of the disease as well as the viral cell tropism. In the most extreme cases, the lack of a particular proviral component would seri-ously impair viral replication and, hence, its capacity to cause disease. The interactions between the hepatitis C virus and miR-122 are a prime illustration of a virus-specific proviral factor (Cabibbo et al., 2012).

HOST RESTRICTION FACTORS

It is essential to highlight that, in addition to proviral factors, specific host proteins can also operate as restriction factors, inhibiting a virus's capacity to multiply and cause disease. Extensive research has been conducted on how TRIM5α belongs to a class of host proteins that include tripartite motifs, several of which have been demonstrated to have immune-regulating or antiviral properties. Research investigating the role of variables in restricting HIV's multiplication in nonhuman primate cells discovered that the TRIM5α molecule in rhesus macaques interacts tightly with the HIV cap-sid at an early stage of the virus's replication process to prevent viral infection. Thus, the macaques are protected against HIV replication and disease. On the other hand, the human TRIM5α molecule does not have an efficient interaction with HIV, making humans more susceptible to HIV infection.

Thus, the existence or lack of suitable restriction factors can significantly affect the sensitivity of the host to virus-induced illness; a strong restriction factor would prevent or reduce viral replication and, thus, host vulnerability to virus-induced illness (McNab et al., 2011).

SHEDDING OF VIRUS

Due to the diversity of viruses, the shedding of infectious viral particles from the infected host into the environment can occur by utilizing many shedding sites. However, the respiratory and alimentary sites are the most frequent sites of shedding. Viral shedding plays a pivotal role in the transmission and spread of infection. Depending on the type of virus and the site of infection, shedding can take place in a number of ways, such as feces, respiratory droplets, genital fluids, blood, skin lesions, and vertical transmission from mother to child. Gastrointestinal viruses, like norovirus, are shed in feces and spread through contaminated surfaces or food. Respiratory viruses, such as influenza and coronaviruses, are typically shed through respiratory secretions during coughing or sneezing. Viruses like HIV and herpes simplex that cause sexually transmitted infections are spread by genital fluids, direct contact with skin sores, and infected blood or blood products (Baron et al., 1996).

REFERENCES

Albrecht, T., Boldogh, I., Fons, M., AbuBakar, S., Deng, C.Z., 1990. Cell activation signals and the pathogenesis of human cytomegalovirus. *Intervirology* 31, 68–75.

Baron, S., Fons, M., Albrecht, T., 1996. Viral pathogenesis. In: Baron, S. (Ed.), *Medical Microbiology*, 4th ed, Galveston, TX, The University of Texas Medical Branch at Galveston.

Cabibbo, G., Maida, M., Genco, C., Antonucci, M., Camma, C., 2012. Causes of and prevention strategies for hepatocellular carcinoma. *Semin Oncol* 39, 374–383.

Li, W., Moore, M.J., Vasilieva, N., Sui, J., Wong, S.K., Berne, M.A., Somasundaran, M., Sullivan, J.L., Luzuriaga, K., Greenough, T.C., Choe, H., Farzan, M., 2003. Angiotensin-converting enzyme 2 is a functional receptor for the SARS coronavirus. *Nature* 426, 450–454.

McNab, F.W., Rajsbaum, R., Stoye, J.P., O'Garra, A., 2011. Tripartite-motif proteins and innate immune regulation. *Curr Opin Immunol* 23, 46–56.

Rose, P.P., Hanna, S.L., Spiridigliozzi, A., Wannissorn, N., Beiting, D.P., Ross, S.R., Hardy, R.W., Bambina, S.A., Heise, M.T., Cherry, S., 2011. Natural resistance-associated macrophage protein is a cellular receptor for sindbis virus in both insect and mammalian hosts. *Cell Host Microbe* 10, 97–104.

Spear, P.G., 2004. Herpes simplex virus: receptors and ligands for cell entry. *Cell Microbiol* 6, 401–410.

9 Double-Stranded DNA Viruses

Kiran Singh, Nilave Ranjan Bora and Sachin Kumar
Department of Biosciences and Bioengineering, Indian
Institute of Technology Guwahati, Guwahati, Assam, India

REPLICATION PROPERTIES OF DOUBLE-STRANDED DNA VIRUSES

The mRNA synthesis by double-stranded DNA (dsDNA) viruses takes place in three steps. In the first step, for transcription initiation, a transcription preinitiation complex is incorporated at the DNA upstream for the binding of the host RNA polymerase. In the second step, RNA polymerase synthesizes mRNA using the negative strand as a template. During the third step, the polyadenylation site acts as a signal for terminating the synthesis. dsDNA viruses use different mechanisms to replicate their genomes, such as the bidirectional replication mechanism, the rolling circle mechanism, and, in some cases, strand displacement and replicative transposition mechanisms are used.

MAJOR FAMILIES OF dsDNA VIRUSES

DNA viruses are classified into nucleocytoplasmic large DNA viruses, *Herpesvirales*, and *Caudovirales*, as tabulated in Table 9.1 (Koonin et al., 2021). Some important families, including *Poxviridae, Adenoviridae, Herpesviridae, Asfarviridae* and *Iridoviridae, Papillomaviridae*, and *Polyomaviridae*, are briefly discussed later in this chapter; others are beyond the scope of this chapter. The typical virus morphology of these important families and their size concerning other viruses are given in Figure 9.1.

POXVIRIDAE

History: The poxviruses are large DNA viruses; the animal infected with poxvirus has characteristic lesions (pockmarks) on the skin. It affects both vertebrates and invertebrates. The history of poxviruses starts with the work done by Edward Jenner. He published his work in 1798, summarizing the clinical signs of pox in cattle (cows) and humans. His work described the clinical symptoms of cattle pox and how human infection protected against smallpox. His work led to vaccination campaigns around the world. But until Pasteur's work, it was not well established. He used the term *vaccine* and *vaccination* (in Latin, *vacca* means "cow") to honor Jenner.

Classification and properties: The family *Poxviridae* has two subdivisions: *Chordopoxvirinae* (vertebrates) and *Entomopoxvirinae* (insects) (Table 9.2). *Chordopoxvirinae* is subdivided into eight genera. *Chordopoxvirinae* is further subdivided into many genera, including *Parapoxvirus, Avipoxvirus*, and *Orthopoxvirus*.

Poxvirus virions are mostly pleomorphic and brick-shaped (220–450 × 140–260 nm), with irregular surface tubules as projections. However, *Parapoxvirus* is ovoid (250–300 × 160–190 nm) with regular surface tubules (Figure 9.1). The genetic material consists of a single linear molecule of dsDNA ranging from 130 to 150 kbp in *Parapoxvirus*, 170–250 kbp in *Orthopoxvirus*, and 300 kbp in *Avipoxvirus*. Poxviruses encode 200 proteins, of which 100 are found in virions. These viral proteins are encoded for transcription and replication activities. Poxvirus also encodes immunomodulating proteins for counteracting the host immune response.

Virus replication: Poxvirus replicates exclusively in the cytoplasm of the host cell. The entry of poxviruses into mammalian cells is facilitated via the attachment of the virions to the cell surface

DOI: 10.1201/9781003369349-9

TABLE 9.1
Universal Taxonomy of dsDNA Viruses

	Family	Virion Morphology	Host
	Papillomaviridae	Icosahedral	Eukarya
	Polyomaviridae	Icosahedral	Eukarya
Nucleocytoplasmic large	*Asfarviridae*	Icosahedral	Eukarya
DNA viruses	*Iridoviridae*	Icosahedral	Eukarya
	Mimiviridae	Icosahedral	Eukarya
	Phycodnaviridae	Icosahedral	Eukarya
	Poxviridae	Oval	Eukarya
	Adenoviridae	Icosahedral	Eukarya
	Ascoviridae	Oval	Eukarya
	Corticoviridae	Icosahedral	Bacteria
	Tectiviridae	Icosahedral	Bacteria
Herpesvirales	*Alloherpesviridae*	Icosahedral	Eukarya
	Herpesviridae	Icosahedral	Eukarya
	Malacoherpesviridae	Icosahedral	Eukarya
Caudovirales	*Myoviridae*	Icosahedral heads, contractile tails	Bacteria, archaea
	Podoviridae	Icosahedral heads, short tails	Bacteria
	Siphoviridae	Icosahedral heads, non-contractile tails	Bacteria, archaea
	Lipothrixviridae	Flexible filaments	Archaea
	Rudiviridae	Stiff rod-shaped	Archaea
	Bicaudaviridae	Spindle-shaped with two tails	Archaea
	Ampullaviridae	Bottle-shaped	Archaea
	Baculoviridae	Rod-shaped	Eukarya
	Fuselloviridae	Spindle-shaped/pleomorphic	Archaea
	Globuloviridae	Spherical, enveloped	Archaea
	Guttaviridae	Droplet-shaped	Archaea
	Nimaviridae	Ovoid/bacilliform with a tail	Eukarya
	Plasmaviridae	Pleomorphic	Bacteria
	Polydnaviridae	Rod-shaped, fusiform	Eukarya

and the fusion/entry event that transports the viral core into the cellular cytoplasm. During the attachment, electrostatic interactions occur between the virion and cell surface moieties, particularly glycosaminoglycans and laminin; moreover, the fusion of the viral membrane with cellular membranes involves a multisubunit viral entry/fusion complex of highly conserved viral proteins (Figure 9.2).

The mRNA transcripts are made from both DNA strands by enzymes within the core, including a virus-encoded multisubunit RNA polymerase. Transcripts are then extruded from the core for translation by host ribosomes. Host macromolecular synthesis is suppressed when early proteins are produced. As a result, basophilic (B-type) inclusions known as "viroplasms" or "virus factories" are created in the cytoplasm of the host cell, where viral replication takes place. The genome has closely spaced open reading frames (ORFs) that can partially overlap and encode proteins. Before DNA replication, one class of early genes is expressed from partially uncoated virions. These genes encode numerous non-structural proteins, such as enzymes for replicating the genome and altering DNA and RNA, as well as proteins responsible for blocking the host response. Also, early genes encode intermediate transcription factors. During DNA replication, intermediate genes are expressed, which are necessary for late gene transcription. These genes encode late transcription factors. Finally, late genes (which mostly encode virion structural proteins) are expressed during the post-replicative phase. The mRNAs are not spliced, have caps, and have been polyadenylated

Double-Stranded DNA Viruses

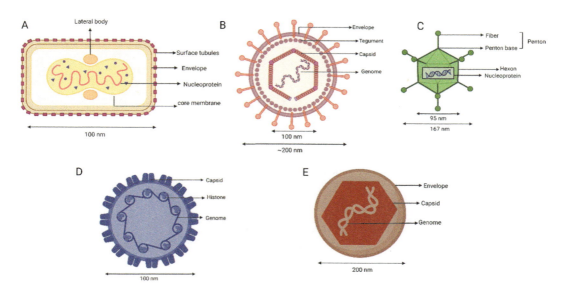

FIGURE 9.1 Representative viruses of dsDNA viruses' family and structure. (A) Vaccinia virus (member of *Poxviridae*), (B) Epstein-Barr virus (member of *Herpesviridae*), (C) adenovirus (member of *Adenoviridae*), (D) papillomavirus (member of *Papillomaviridae*), and (E) frog virus 3 (member of *Iridoviridae*). (Adapted and created from Biorender, 2023; Maclachlan and Dubovi, 2010.)

TABLE 9.2
Taxonomical Classification of Poxviruses and Their Geographical Distribution

Poxviruses: Major Host and Geographical Distribution

Genus	Common Examples	Host	Geographical Distribution
Orthopoxvirus	Smallpox (Variola)	Humans	Eradicated globally worldwide
	Vaccinia virus	Numerous	
	Cowpox virus	Numerous	Europe, Asia
	Camel pox virus	Camels	Asia, Africa
	Monkeypox virus	Numerous	Western and Central Africa
Capripoxvirus	Sheeppox virus	Sheep, goats	Africa, Asia
	Goatpox virus	Goats, sheep	Africa, Asia
	Lumpy skin disease virus	Cattle, Cape buffalo	Africa
Suipoxvirus	Swinepox virus	Swine	Worldwide
Leporipoxvirus	Myxoma virus	Rabbits	America, Europe, Australia
Molluscipoxvirus	Molluscum contagiosum virus	Primates including humans, birds, kangaroos	Worldwide
Yatapoxvirus	Yabapox virus and Tanapox virus	Monkeys, humans	West Africa
Avipoxvirus	Fowlpox virus, canarypox, sparrow pox viruses	Chicken, turkeys, etc.	Worldwide
Parapoxvirus	Orf virus, pseudocowpox virus	Sheep, goats, humans	Worldwide

at their 3′ termini. Many intermediate, late, and some early mRNAs feature 5′-poly(A) tails before the encoded mRNA. Although early protein synthesis typically shuts down as late gene expression takes over, some genes can still be produced through promoters with both early and late activity. Posttranslational modifications to proteins include phosphorylation, glycosylation, ribosylation,

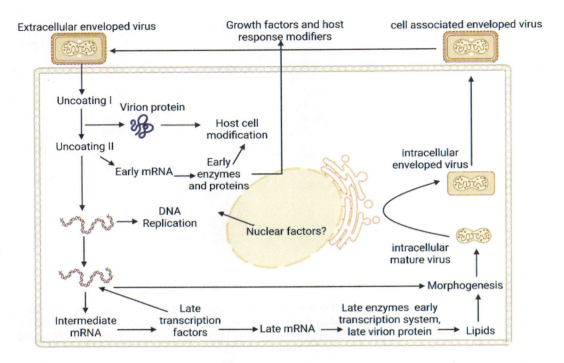

FIGURE 9.2 Schematic representation of the replication cycle of the vaccinia virus (poxvirus). (Adapted and created from Biorender, 2023; Maclachlan and Dubovi, 2010.)

sulfation, acylation, palmitoylation, and myristoylation. Other posttranslational modifications include proteolytic cleavage. Virion formation requires the proteolytic cleavage of late proteins.

It appears that viral enzymes are primarily responsible for the replication of the DNA genome. A single-stranded nick is introduced to begin DNA replication near the terminal hairpins. The hairpin is unfolded and copied to the terminus. Nascent DNA can be extended along the entire genome length, through the opposing hairpin, and back along the opposing strand to generate a concatemeric product. The two strands are split at the terminal, and the hairpins re-join. At least three virus-encoded enzymes—a DNase, topoisomerase I, and a Holliday junction resolvase—are involved in the resolution of the concatemers for packaging. After DNA replication and the expression of early, intermediate, and late genes, virus morphogenesis starts. Creating crescent-shaped membrane structures between the endoplasmic reticulum and the trans-Golgi network marks the beginning of particle assembly. Concatemeric DNA that has been replicated is divided into unit genomes and packaged to create virion particles, eventually developing into fully infectious intracellular mature virus (IMV). To create an intracellular enveloped virus (IEV), some MV acquire a second double layer of the internal membrane (derived from early endosomes or the trans-Golgi network) containing distinctive viral proteins. These IEVs are moved to the cell's periphery by joining the cellular microtubule network, fusing with the plasma membrane to release EV (cell-associated EV and extracellular EV). While both MV and EV are contagious, their surface antigens differ, and during infection, the two virion types likely bind to various cellular receptors before being taken up by the mechanisms mentioned earlier. All infectious virions have a nucleoprotein complex composed of protein and viral DNA. The nucleoprotein core complex, lateral bodies, and an all-encompassing surface membrane comprise the MV (IMV). The core wall of the vaccinia virus has a typical component structure. Negative staining within the vaccinia virion reveals that the core adopts a biconcave form, seemingly due to the massive lateral bodies.

Double-Stranded DNA Viruses

HERPESVIRIDAE

History: Hippocrates was the first to describe lesions that might be caused by herpes simplex virus (HSV), but the clinical conditions have been described in more detail over the past three centuries. It was not until 1893 that Vidal recognized the human transmission of herpes simplex infection from one individual to another. Histopathologic studies characterized the multinucleated giant cells associated with herpesvirus infection. In 1919, Lowenstein confirmed experimentally the infectious nature of HSV that Shakespeare had only suspected. In the 1920s and 1930s, the natural history of HSV was widely studied, and it was found to infect the skin and the central nervous system. In the 1930s, host immune responses to HSV were thoroughly examined, and the property of HSV known as latency was characterized. By the 1940s and 1950s, research abounded on the many diseases caused by HSV. More recent research has focused on antiviral research, differences between HSV strains, and using HSV vectors in vaccines.

Classification and properties: Classification of herpesviruses is complex, but they are divided into *Herpesviridae* (herpesviruses of mammals), *Alloherpesviridae* (herpesviruses of fish and frogs), and *Malacoherpesviridae* (herpesviruses of oysters). *Herpesviridae* is subdivided into subfamilies: *Alphaherpesvirinae*, *Betaherpesvirinae*, and *Gammaherpesvirinae*. Typical herpesvirus virions are enveloped, including a core and tegument. The capsid is composed of 162 capsomers (150 hexons and 12 pentons). The genome is linear double-stranded DNA variably sized between 125 and 290 kbp; variability occurs in composition, size, and organization.

Virus replication: Herpesvirus replication is most studied with human herpes simplex virus 1; the overall replication of the virus is the same, but considerable variation is found. The virus migrates into the host cell via receptor-mediated binding of viral glycoprotein spike proteins. This allows the virus to fuse with the cell plasma membrane, and the nucleocapsid enters the host cytoplasm. Once viral proteins are freed from the nucleocapsid, they hamper the host machinery and start replication. Then host RNA polymerase II synthesizes three classes of mRNA: α, β, and γ. Early αRNA is transcribed into mRNA, which further translates α proteins. The α proteins initiate the transcription of βRNA and inhibit its further transcription. The viral DNA utilizes these proteins with some of the host proteins to replicate its genome. During the late phase of replication, the virus transcribes γRNA, which translates structural proteins. Viral DNA, after replication, is coiled into immature capsids. The virions are encapsidated into nucleocapsids while budding through the nuclear membrane. Newly formed virions accumulate in the vacuoles of the cytoplasm before being released by exocytosis into the extracellular milieu.

PAPILLOMAVIRIDAE AND POLYOMAVIRIDAE

Papillomaviridae

Classification and properties: The double-stranded DNA genomes of the *Papillomaviridae* family of small, non-enveloped viruses range in size from 5,748 bp to 8,607 bp. They are categorized based on pairwise nucleotide sequence identity throughout the L1 open reading frame. Members of *Papillomaviridae* have been found in fish, reptiles, birds, and mammals; they primarily infect mucosal and keratinized epithelia. The virions of the papillomavirus lack an envelope. The major capsid protein, L1, has 360 copies organized as 72 pentamers, and the minor capsid protein, L2, has 12 molecules (Figure 9.1). Virus-like particles can self-assemble when recombinant L1 is expressed, either with or without L2. The viral circular dsDNA is packaged in a single copy by each capsid. Core histone proteins are connected to the packaged viral DNA.

Virus replication: The circular virus's genome is translated from just one DNA strand. Three functional areas can be identified in the viral genome. Virus proteins crucial in transcription, replication, and control of the cellular environment are encoded in the early time-point. The capsid proteins L1 and L2 are encoded in the late region. Between the L1 and E6 ORFs, the upstream regulatory region (URR or LCR) houses the viral replication origin and binding sites for cellular and viral transcription factors. There are three main phases of replication in the viral replication cycle. The viral E1 and E2 replication proteins support initial, confined viral DNA amplification. The viral E1 helicase is recruited when the viral E2 protein binds to its binding sites in the viral origin of replication, enabling viral replication. The viral genome is kept at a relatively low but constant copy number in the proliferating cells of a clonally enlarged population of infected cells during maintenance replication, which follows this first burst of replication. Finally, genome amplification and the production of virions occur as an infected cell completes cellular differentiation. The virus must create an S-phase-like condition in differentiated cells during maintenance replication. The viral E6 and E7 proteins hijack the cellular environment through various protein–protein interactions, enabling viral reproduction in differentiated cells. The viral life cycle's maintenance phase might persist for months or years. The viral E2 protein controls replication and plays a crucial role in maintenance by ensuring that the viral genomes are accurately distributed into the daughter cells. The viral DNA has been amplified to a high copy number in the upper layers of differentiated epithelia. To form particles encasing the viral DNA, the viral capsid proteins self-assemble. Infectious virions are released as the cells slip into the environment, concluding the viral life cycle.

Polyomaviridae

Classification and properties: *Polyomaviridae* is a family of small, non-enveloped viruses with dsDNA genomes of approximately 5,000 bp. Phylogenetic relationships among polyomaviruses, based on the amino acid sequence of the viral protein large tumor antigen, have resulted in the delineation of 117 species, 112 of which have been assigned to 6 genera: *Alphapolyomavirus*, *Betapolyomavirus*, *Gammapolyomavirus*, *Deltapolyomavirus*, *Epsilonpolyomavirus*, and *Zetapolyomavirus*. The members of these genera can infect mammals and birds, and polyomavirus genomes have recently been detected in fish. Each family member has a restricted host range. Some members are known human and veterinary pathogens causing symptomatic infection or cancer in their natural host. Clinical symptoms are observed primarily in immunocompromised patients. The genome of most polyomaviruses is approximately 5,000 bp (Figure 9.1). The genomes of members of the recognized species vary from 3,962 bp to 7,369 bp. Of the known human polyomaviruses, the Merkel cell polyomavirus has the largest genome (5,387 bp), and the Saint Louis polyomavirus has the smallest (4,776 bp).

Virus replication: The genome contains two distinct transcriptional regions: the early and the late regions, depending on the stage of infection. A non-coding control region (NCCR) separates the early and late regions. Transcription of the early region results in a single precursor mRNA from which different transcripts are generated through alternative splicing. The major translational products generated from these spliced mRNAs are the regulatory proteins LTAg and STAg. Several polyomaviruses express additional early proteins, or their genomes encode additional putative early proteins. The late region is transcribed from the complementary strand and in the opposite direction from the early region. The late region codes for at least two late proteins, VP1 and VP2, translated from different mRNAs due to alternative splicing. For most polyomaviruses, a third structural protein (VP3) is generated from the same transcript as the VP2 protein using an internal, in-frame start codon. For SV40, a VP4 was identified in the same ORF as VP2/VP3 and was found necessary for lysis-infected cells. Some polyomaviruses have a short ORF upstream of the start codon of VP1, the agno gene, that encodes a hydrophobic protein known as agnoprotein. The VP4 protein of bird polyomaviruses is derived from an additional ORF upstream of the VP2-encoding late mRNA. Some mammalian polyomavirus lineages result from ancient recombination events between early and late regions of the genome.

The NCCR, which is located between the early and late regions, directs the transcription of the early and late genes. The NCCR shows the highest sequence variability among polyomaviruses. However, conserved LTAg-binding motifs (5'-GRGGC-3') are present throughout the family. The early promoter has a TATA box, whereas the late promoter lacks such a motif. The NCCR contains multiple binding sites for cellular transcription factors. LTAg is involved in the switch from early

Double-Stranded DNA Viruses

to late viral gene expression. At low concentrations of LTAg, this protein will occupy high-affinity LTAg-binding motifs and stimulate early transcription and viral DNA replication. Later in infection, LTAg will also bind to low-affinity binding motifs as LTAg concentration increases. These motifs are located downstream from the TATA box. As a result, LTAg prevents early transcription by blocking the passage of the RNA polymerase II complex. LTAg is also involved in the switch to late transcription and facilitates late transcription by recruiting transcription factors. For SV40, it was shown that a cellular repressor present in low concentration prevents late transcription. As the viral genome replicates, more viral genome copies are produced, and the repressor is depleted, allowing late transcription. Viral DNA transcribed during the late phase of infection seems nucleosome-free in the NCCR, allowing more active transcription.

ASFARVIRIDAE AND *IRIDOVIRIDAE*

Asfarviridae

Classification and properties: The African swine fever virus contains linear dsDNA genomes of 170–194 kbp and is a member of the *Asfarviridae* family (Table 9.1). The virion structure includes an icosahedral capsid, an internal core, an interior lipid layer, and an external lipid envelope. Hemorrhagic fever and mortality are common in domestic pigs and wild boar infections. Contact, ingestion, and ticks from the genus *Ornithodoros* are all potential methods of infection transmission. Africa is an endemic zone, where warthogs and bush pigs are reservoirs. The virus traveled from West Africa to the Iberian Peninsula, where it remained endemic for 30 years before being eventually eradicated. The virus once more made its way from Africa through the Caucasus into Europe in 2007, and outbreaks occurred in the Russian Federation and other nearby nations, including the Baltic states, Poland, the Czech Republic, and Romania.

Virus replication: The mononuclear phagocytic system is the main cell type infected by the virus. Virus replication occurs in vitro in macrophages and endothelial cells adapted to the cell culture system. The primary method by which the virus enters cells is clathrin- and dynamin-dependent receptor-mediated endocytosis, either in conjunction with the macrophages' inherent ability to enter cells through macropinocytosis or as a separate entry mechanism (Figure 9.3). ASFV capsid breakdown occurs at the endosomal lumen's acidic pH, and ASFV uncoating depends on host factors present during its endosomal transit. The viral protein pE248R and the stability of cholesterol flow are required to fuse the virion's internal membrane with the endosomal membrane. Viral cores emerge from endosomes following membrane fusion and capsid breakdown. The ubiquitin–proteasome system finally degrades the viral cores, releasing the viral DNA and allowing it to begin replicating. Using enzymes and substances in the virus core, early mRNA synthesis starts as soon as the virus enters the cell. In perinuclear factories, also called viral factories, where the virus is carried along with the microtubular motor light chain dynein, viral DNA replication and assembly occur. Early after infection, virus DNA is found in the nucleus, indicating that nuclear enzymes may play a part in the early phases of DNA replication. After six hours of infection, DNA concatemers of the head-to-head virus are found in the cytoplasm and are assumed to be replicative intermediates. Similar to viruses in the *Poxviridae* family, the cytoplasmic replication of DNA follows a similar method.

Iridoviridae

Classification and properties: Members of the family *Iridoviridae* comprise a collection of large icosahedral, dsDNA-containing viruses that are classified into two subfamilies: *Alphairidovirinae* and *Betairidovirinae* (Table 9.1). The former comprises three genera (*Ranavirus*, *Megalocytivirus*, and *Lymphocystivirus*) whose members primarily infect ectothermic vertebrates such as bony fish, amphibians, and reptiles. In contrast, the latter contains three genera (*Iridovirus*, *Chloriridovirus*, and *Decapodiridovirus*) whose members infect mainly invertebrates such as insects and crustaceans.

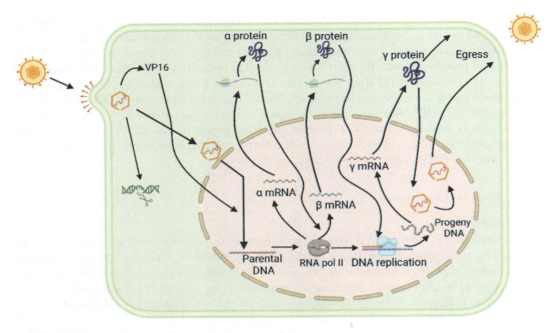

FIGURE 9.3 Schematic representation of the replication cycle of the typical herpesvirus. (Adapted and created from Biorender, 2023; Maclachlan and Dubovi, 2010.)

Viral macromolecular synthesis involves nuclear and cytoplasmic compartments, and virion assembly occurs in the cytoplasm within morphologically distinct viral assembly sites. Mature, non-enveloped but otherwise infectious virions may remain within the cytoplasm, from which they are released by cell lysis, or virions may acquire an envelope by budding from the plasma membrane.

Virus replication: Iridovirid genomes have a unique component and a variable amount of terminal redundancy. The unique component ranges from 103 to 220 kbp, whereas the actual genome content is 5–50% larger due to terminal redundancy and varies depending on the species. Except for LCDV1 (103 kbp), genomes for members of the *Lymphocystivirus*, *Iridovirus*, *Chloriridovirus*, and *Decapodiridovirus* genera are larger (163–220 kbp) than those for members of the *Megalocytivirus* and *Ranavirus* genera (103–140 kbp). Except for SGIV, members of species within the genus *Ranavirus* tend to show marked sequence co-linearity and a high cytosine methylation level. However, genome inversions and deletions have been noted even within this genus. Putative ORFs are found on both strands of the viral genome, but the overlap of coding regions is rare. Moreover, gene density is high, intergenic regions are generally short, and genomes contain repetitive sequences. In contrast to eukaryotes, iridovirid genes lack introns, and viral mRNAs lack poly[A] tails. However, like eukaryotes, biochemical and in silico evidence suggests the existence of viral microRNAs (miRNAs) that modulate viral gene expression.

Iridovirids employ a novel replication strategy involving nuclear and cytoplasmic stages that have been elucidated primarily through the study of FV3, the type species of the genus *Ranavirus*. Although the precise mechanism of virion entry remains to be elucidated, multiple routes have been implicated, including receptor-mediated endocytosis by enveloped particles, uncoating of naked virions at the plasma membrane, pinocytosis, and caveola-dependent endocytosis. Following uncoating, viral DNA cores enter the nucleus, where first-stage DNA synthesis and the synthesis of immediate early (IE) and delayed early (DE) viral transcripts take place. In a poorly understood process, one or more virion-associated proteins act as transcriptional transactivators and redirect host RNA polymerase II to synthesize IE viral mRNAs using the methylated viral genome as a template. Gene products encoded by IE (and perhaps DE) viral transcripts include regulatory and

Double-Stranded DNA Viruses

catalytic proteins. One of these gene products, the viral DNA polymerase, catalyzes the first stage of viral DNA synthesis. The parental viral genome serves as the template, and progeny DNA is synthesized to genome-length at most twice. Newly synthesized viral DNA may serve as the template for additional rounds of DNA replication or early transcription, or viral DNA may be transported to the cytoplasm, where the second stage of viral DNA synthesis occurs. Second-stage DNA synthesis results in the formation of large, branched concatemers. Viral DNA methylation also occurs in the cytoplasm, and although its precise role is uncertain, it is thought to protect viral DNA from virus-mediated endonucleolytic attack. Late (L) viral transcripts are synthesized in the cytoplasm and catalyzed by virus-encoded homologs of RNA polymerase II. Like other large DNA viruses, full L gene expression requires prior DNA synthesis. Thus, temperature-sensitive mutants that fail to synthesize viral DNA at the non-permissive temperature and infected cells incubated in the presence of inhibitors of DNA synthesis display markedly reduced levels of late viral transcripts and proteins. Virion formation occurs in the cytoplasm within morphologically distinct regions termed virus assembly sites. Assembly sites form in cells treated with an antisense morpholino oligonucleotide (asMO) targeted against vPOL-IIα, indicating that assembly site formation does not require the synthesis of late viral gene products. Concatemeric viral DNA is thought to be packaged into virions via a "headful" mechanism that generates circularly permuted and terminally redundant genomes. Following assembly, virions accumulate in the cytoplasm within paracrystalline arrays or acquire an envelope by budding from the plasma membrane (Figure 9.4). In the case of most vertebrate iridoviruses, the majority of virions remain cell-associated.

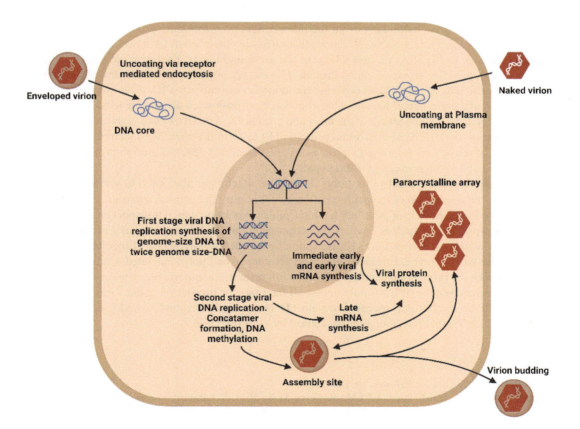

FIGURE 9.4 Schematic representation of the typical frog virus 3 replication cycle. (Adapted and created from Biorender, 2023; Maclachlan and Dubovi, 2010.)

ADENOVIRIDAE

Classification and properties: The *Adenoviridae* virus family has non-enveloped, icosahedral virions with linear dsDNA genomes ranging from 25 to 48 kb. Family members are divided into six genera, infecting various vertebrate hosts (including everything from fish to humans). *Mastadenovirus* strains infect mammals, *Aviadenovirus* strains infect birds, *Ichtadenovirus*'s single fish adenovirus infects fish, and *Testadenovirus* strains infect turtles. Members of the *Siadenovirus* family infect birds, frogs, marsupials, and tortoises, while members of the *Atadenovirus* family infect squamate reptiles, birds, ruminants, and tortoises. Adenovirus (AdV) virions are non-enveloped 90 nm diameter particles. The typical icosahedral capsid has a pseudo T = 25 triangulation number. It comprises 240 non-vertex capsomers (hexons), which are 8–10 nm in diameter, and 12 vertices (penton bases), from which fibers protrude. The 240 hexons have two unique parts: a triangular top with three "towers" and a pseudohexagonal base with a central cavity. They are created by interacting with three identical proteins. Hexon bases are closely packed to create a protein shell, shielding the interior parts. The 12 penton bases are securely bonded to the fibers, which are homotrimers of protein IV and are each generated by the interaction of five copies of protein III. The fibers have three structural domains: the distal knob, the shaft of a particular length, and the tail, which attaches to the penton base. The fibers range in length from 9 to 77.5 nm. The vertices of human adenoviruses 40 and 41 alternately include fibers of two distinct lengths. Many simian adenoviruses (SAdVs) and human adenovirus 52 contain fibers of two lengths. Some SAdVs even have three different fibers. Multiple fibers may be inserted at a single vertex in *Aviadenovirus* and *Atadenovirus* members. Protein IX of the genus *Mastadenovirus*, which has 12 copies per facet and is found on the exterior of the capsid, binds the hexon proteins. Protein IX also connects two facets across the icosahedral edge in human adenoviruses. Individuals from the other genera do not contain protein IX. Despite having a distantly related structure to IX, the atadenovirus-specific protein LH3 performs structurally similar functions to IX. IIIa, VI, and VIII proteins are found inside. Each penton base has a ring-like arrangement of five IIIa monomers. Protein VI is not organized in an icosahedral pattern and is found below the hexons. Each facet contains six monomers of protein VIII, three wedged between protein IIIa and the peripheral hexons. The other three are positioned around the icosahedral three-fold symmetry axis, helping stabilize the facet and protein IX. The DNA genome is complexed with the four proteins V, VII, X, and terminal protein (TP) to form the core. Mastadenoviruses are the only source of protein V. Adenovirus protease/AVP/adenain/L3 23K maturation protease is packaged and coupled to the genome in a few copies.

Virus replication: A typical virus enters the cell by attaching a fiber knob to various receptors, governed by the interaction of the penton base with cellular αv integrins. Virions are released from the endosomes with the help of protein VI, which guides dynein-mediated transport on microtubules to the nuclear pores. The nucleus is where virus transcription, DNA replication, and virion assembly occur after the uncoated virus core. Transcription by host RNA polymerase II involves both DNA strands of the virus genome. Primary transcripts are capped and polyadenylated. Complex splicing patterns govern the production of mRNA families. In primate adenoviruses, virus-associated RNA genes transcribed by cellular RNA polymerase III facilitate the translation of late virus mRNAs and block the cellular interferon response. Early host DNA synthesis termination and later host mRNA and polypeptide synthesis termination are caused by viral infection.

BIBLIOGRAPHY

Biorender, 2023. *Created with Biorender*. Biorender.

Koonin, E.V., Krupovic, M., Agol, V.I.J.M., Reviews, M.B., 2021. The Baltimore classification of viruses 50 years later: how does it stand in the light of virus evolution? 85. doi:10.1128/mmbr. 00053-00021.

Maclachlan, N.J., Dubovi, E.J., 2010. *Fenner's Veterinary Virology*. Academic Press.

10 Single-Stranded DNA Viruses

Anusha Sairavi[1] and Yoya Vashi[2]
[1]Department of Molecular and Medical Genetics, Oregon Health and Science University, Portland, OR, USA
[2]Department of Surgery, City of Hope National Medical Center, Duarte, CA, USA

INTRODUCTION

A significant component of the Earth's virome includes single-stranded (ss) DNA viruses. ssDNA viruses represent a diverse group of medically, ecologically, and economically important pathogens infecting hosts from all three domains of cellular life: archaea, bacteria, and eukaryotic organisms. The abundant and widespread nature of ssDNA viruses in diverse environments indicates that they are competitive members of the global virome. The International Committee on Taxonomy of Viruses (ICTV) has established 23 ssDNA virus families, out of which 21 contain circular genomes. *Parvoviridae* and *Bidnaviridae* are the only exceptions that contain linear genomes. Seventeen of these 21 circular ssDNA virus families infect eukaryotic organisms. The families *Alphasatellitidae*, *Anelloviridae*, *Finnlakeviridae*, *Spiraviridae*, and *Tolecusatellitidae* are established and proposed ssDNA virus taxa that have not yet been assigned to any realm.

TAXONOMY AND CLASSIFICATION

ssDNA viruses fall under the second Baltimore classification (BCII), and this classification remains stable even after several years with minor exceptions (Koonin et al., 2021). BCII classification can be defined as "minimal" viruses with ssDNA genomes that often only encode for a protein involved in replication and a structural protein that forms the capsid structure. At the time of BCII classification, ssDNA viruses carried either (+)sense genomes or both sense genomes. However, this has since changed with the discovery of *Anelloviridae* that carry (–)sense genomes. The majority of the viruses under BCII have single-stranded genomes except for *Pleolipoviridae*, which can carry either single- or double-stranded DNA, and *Bacilladnaviridae*, which are majorly single-stranded but have short double-stranded regions. Thus, technically, both pleolipoviruses and bacilladnaviruses can also be considered representatives of two different BCs, namely, BCI and BCII. Also, most viruses under this classification have circular genomes except for *Bidnaviridae* and *Parvoviridae*. This group of circular virus families with single-stranded DNA encoding for replication-associated protein (Rep) has been collectively called circular Rep-encoding Single-stranded DNA (CRESS) DNA viruses, which is discussed later. The ssDNA families and their key features are listed in Table 10.1.

GENOME ARCHITECTURE

Most ssDNA viruses possess circular genomes except for *Bidnaviridae* and *Parvoviridae*, which have linear genomes and infect a wide variety of prokaryotic and eukaryotic hosts (Zhao et al., 2019). Most ssDNA viruses are considered minimalistic viruses, with genome lengths as short as 1 kb and harboring only two genes. Such simplicity of ssDNA viruses exemplifies their essence of being a virus and makes them an attractive model for investigating virus origins and evolution.

DOI: 10.1201/9781003369349-10

TABLE 10.1

Features of ssDNA Virus Families

Host	Virus Taxonomy	Subfamily	Genera	Virus Morphology	Genome Morphology	Genome Size (kb)	Replication
Bacteria	*Finnlakeviridae*	—	1	Icosahedral	Circular	9.2	Possibly rolling circle
	Inoviridae	—	25	Filamentous	Circular	5.5–10.6	Rolling circle
	Microviridae	2	7	Icosahedral	Circular	4.4–6.1	Rolling circle
	Paulinoviridae	—	2	Filamentous	Circular	5.6–5.8	Rolling circle
	Plectroviridae	—	4	Rods	Circular	4.5–8.3	Rolling circle
Archaea	*Pleolipoviridae*		3	Pseudo-spherical and pleomorphic	Circular	7–17	Possibly rolling circle replication for circular genomes; protein primed replication for linear genomes
	Spiraviridae		1	Helical with protrusions from both termini	Circular	24.9	Non-lytic, chronic infection
Eukarya	*Alphasatellitidae*	2	11	Icosahedral	Circular	1.1–1.4	Rolling circle
	Anelloviridae	—	14	Icosahedral	Circular	2–3.9	Possibly rolling circle
	Bacilladnaviridae	—	3	Icosahedral	Circular, a segment of linear genome present	4.5–6	Rolling circle
	Bidnaviridae	—	1	Icosahedral	Linear, bipartite	6–6.5	DNA polB primed extension
	Circoviridae	—	2	Icosahedral	Circular	1.7–3.8	Rolling circle
	Geminiviridae	—	14	Icosahedral, twinned	Circular	2.5–5.2	Rolling circle
	Genomoviridae	—	10	Icosahedral	Circular	2.17	Rolling circle
	Metaxyviridae	—	1	Icosahedral	Circular	1.3	Possibly rolling circle
	Nanoviridae	—	2	Icosahedral	Circular, multipartite	0.9–1.1	Rolling circle
	Naryaviridae	—	4	Icosahedral	Circular	1.8–2.7	Rolling circle
	Nenyaviridae	—		Icosahedral	Circular	1.7–3	Rolling circle
	Parvoviridae	3	26	Icosahedral	Linear	4–6	Rolling hairpin
	Redondoviridae	—	1	Icosahedral	Circular	3–3.1	Possibly rolling circle
	Smacoviridae	—	12	Icosahedral	Circular	2.3–3	Rolling circle
	Tolecusatellitidae	—	2	Icosahedral	Circular	0.7–1.35	Rolling circle
	Vilyaviridae	—	12	Icosahedral	Circular	1.9–2.2	Rolling circle

Single-Stranded DNA Viruses

TABLE 10.2

Known ORFs and Proteins Encoded by ssDNA Virus Families

Virus Taxonomy	Number of Proteins Encoded	Major Proteins Encoded
Alphasatellitidae	1	Replication proteins
Anelloviridae	2 major and other predicted	Structural and replication proteins, proteins with phosphatase activity, proteins involved in suppression of the NF-κB pathway
Bacilladnaviridae	3	Structural and replication proteins
Bidnaviridae	4 major and other predicted	Type B DNA polymerase, structural and other accessory proteins
Circoviridae	2 major	Structural and replication proteins
Finnlakeviridae	16 predicted	Structural and other accessory proteins
Geminiviridae	2 major	Structural and replication proteins
Genomoviridae	2 major	Structural and replication proteins
Inoviridae	7–15	Structural and replication proteins, morphogenesis proteins, integration proteins
Metaxyviridae	6	Structural and replication proteins and other accessory proteins
Microviridae	11	Structural and replication proteins, spike proteins, and other accessory proteins
Nanoviridae	Multipartite, 6–8 segments, monocistronic	Structural and replication proteins, proteins related to cell cycle and nuclear shuttling
Naryaviridae	2	Structural and replication proteins
Nenyaviridae	2	Structural and replication proteins
Parvoviridae	2+	Structural, replication, and other accessory proteins
Paulinoviridae	8–10	Structural and replication proteins, morphogenesis proteins, and other accessory proteins
Plectroviridae	4–10	Structural, replication, and other accessory proteins
Pleolipoviridae	8–16	Structural and replication proteins, spike proteins, and other accessory proteins
Redondoviridae	3	Structural, replication, and an unknown protein
Smacoviridae	2 major	Structural and replication proteins
Spiraviridae	57 predicted	Structural and other non-structural proteins
Tolecusatellitidae	1	Replication protein
Vilyaviridae	2	Structural and replication proteins

However, some families package longer genomes up to 25 kb and can encode up to 10 ORFs (open reading frames). In general, like all other viruses, ssDNA viruses code for capsid proteins (CPs), replication initiation proteins (Rep), and other accessory proteins involved in the viral life cycle. The Rep shows a great degree of conservation, whereas the CP is highly variable. The arrangement of Rep/CP ORFs either diverges or converges toward the origin of replication (Malathi and Renuka Devi, 2019). A summary of the ORFs encoded by various ssDNA virus families and the major proteins coded by them is provided in Table 10.2.

VIRUS REPLICATION

Virus replication is the process through which the virus forms new virions, which are eventually packaged into the capsids. Three modes of replication can be found in ssDNA viruses: rolling circle, rolling hairpin, and protein-primed DNA polymerase-mediated replication. Some families have replication modes that have not been experimentally verified but are most likely to fall within the three modes. The first two modes of replication—rolling circle and rolling hairpin—are mediated by endonucleases called Rep proteins, while the third mode is mediated through protein-primed DNA polymerase.

1) *Rolling circle*: A commonly used mode of replication by all circular viruses, the process is initiated by the Rep protein that binds to the genome at a specific site called the origin of replication (*v-ori*) and nicks it. The nick generates a 3′ OH free end, which is then extended by the host DNA polymerase to synthesize a new strand using the circular strand as a template. The new strands are then encapsidated to form new virions. The old strand is displaced to form concatemerized linear strands (Figure 10.1).

2) *Rolling hairpin*: The mode of replication used by linear DNA of the family *Parvoviridae* is named such because of the replication at the inverted terminal repeats forming the hairpin structure. Mechanistically like the rolling circle, the Rep protein binds to the Rep binding site at the 3′ hairpin and nicks it, creating a free 3′ OH end. The 3′ OH end is then extended by DNA polymerase to create a new strand that is eventually displaced and packaged to form new virions (Figure 10.2).

3) *Protein-primed DNA polymerase*: This form of replication is exhibited by the virus members of the family *Bidnaviridae*. Bidnaviruses encode a type B DNA polymerase and are predicted to replicate their genome via the protein-primed mechanism reminiscent of that

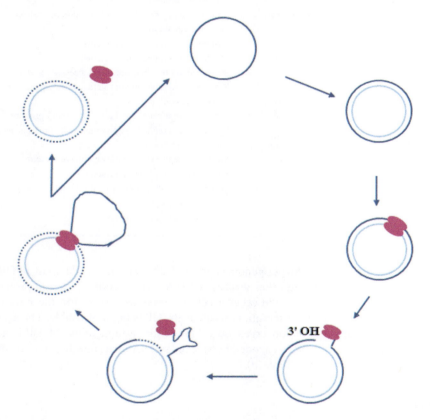

FIGURE 10.1 Schematic representation of rolling circle replication (RCR) in circular ssDNA viruses. The encapsidated ssDNA released into the host cell enters the nucleus and is converted to dsDNA by host enzymes. Rep is then expressed, which binds and nicks the virion strand origin of replication. As a new strand is synthesized, the old strand is displaced. When replication is completed, Rep joins the nicks and releases the initial strand from the replication complex.

Single-Stranded DNA Viruses

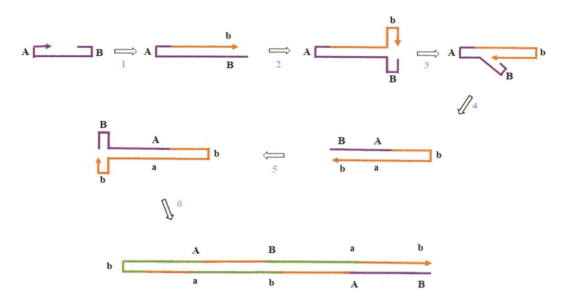

FIGURE 10.2 Schematic representation of rolling hairpin replication (RHR) in circular ssDNA viruses. Upon infection of the host cell, the 3′ end of the linear ssDNA serves as a primer, and host enzymes convert ssDNA to dsDNA. The new strand is synthesized by host DNA polymerase. Instead of simply ending at this point, the replication complex switches from the parental strand to the identical sequence on the newly synthesized strand, and replication continues to produce a dsDNA molecule with paired hairpins at one end. This process continues back and forth, producing genomic concatemers within which the palindromic genome ends are replicated half as frequently as the coding regions. The double-stranded molecules are further processed to release genome lengths of ssDNA molecules that are ready for packaging.

characterized in adenoviruses. A terminal protein encoded by the viral genome heterodimerizes with the DNA polymerase encoded by the viral genome. This complex then recognizes the origins of replication on both ends of the genome through terminal proteins that were bound during previous rounds of replication. After this, the DNA polymerase releases itself from the complex and extends to form new strands.

ORIGIN AND EVOLUTION

Over the years, high diversity of ssDNA viruses has been found in various habitats making up a large part of Earth's virome. The family *Parvoviridae* was the first to be discovered in 1973, with several families discovered after that. The miniscule size of these ssDNA viruses has led to significant undercounting and hence varied discovery rates. Metagenomics data has laid an extensive foundation into the origin and evolution of ssDNA viruses, particularly viruses infecting eukaryotes. Most of the ssDNA viruses replicate via rolling circle replication (rolling hairpin in the case of *Parvoviridae*) that is mediated by an endonuclease, Rep protein. Surprisingly, the Rep of eukaryotic viruses shows little similarity to the Rep encoded by the prokaryotic viruses. The Rep of eukaryotic viruses shows two domain organizations: His hydrophobic His (HUH) endonuclease and superfamily 3 helicase (S3H) domain, which shows sequence similarity to several bacterial plasmids. In fact, the Rep of different eukaryotic viruses shows homology to bacterial plasmids than they show to each other. Similarly, the prokaryotic Rep, although missing the S3H domain, also shows homology to bacterial plasmids (Kazlauskas et al., 2019). Although the Rep of different families of ssDNA viruses is mechanistically similar in their activity, they have been acquired independently and from various sources. A striking deviation from the Rep-mediated replication is the family *Bidnaviridae* that encodes for protein-primed family B DNA polymerase instead of Rep. Evolutionary analysis of *Bidnaviridae* shows that the polymerase was acquired from virus-like DNA transposons along with acquiring other accessory genes from insect viruses (Krupovic and Koonin, 2014). *Bidnaviridae* was once grouped under the family *Parvoviridae* but has since become its own family and seems to have evolved from its ancestor along with other recombination events exemplifying the diversity that ssDNA viruses possess. In contrast, the capsid proteins (CPs) of most of the ssDNA viruses have a characteristic jelly roll fold that was acquired from various plant and animal (+)RNA viruses and are strikingly different from each other (Diemer and Stedman, 2012; Kazlauskas et al., 2019). Given the origin of Rep from bacterial plasmids and CPs from (+)RNA viruses, a recombination event between RNA and DNA replicons must have occurred on several independent occasions.

EUKARYOTIC CIRCULAR REP-ENCODING SINGLE-STRANDED DNA (CRESS DNA) VIRUSES

CRESS DNA viruses refer to a group of ssDNA viruses encoding a Rep that appears to be descended from a common ancestor and are grouped together in the phylum *Cressdnaviricota*. Another characteristic feature of CRESS DNA viruses is the presence of an N-terminal rolling circle replication initiation endonuclease domain of the HUH superfamily and a C-terminal S3H domain. Most ssDNA viruses are CRESS DNA viruses and display global environmental distribution and infect diverse eukaryotic hosts, including animals, plants, and fungi. Prior to 2015, only three families of CRESS DNA viruses, namely, *Geminiviridae*, *Circoviridae*, and *Nanoviridae*, were recognized by ICTV, which later increased with the addition of *Bacilladnaviridae*, *Genomoviridae*, *Redondoviridae*, *Metaxyviridae*, and *Smacoviridae*. Very recently, another three families (*Naryaviridae*, *Nenyaviridae*, and *Vilyaviridae*) of CRESS DNA viruses have been created (Krupovic and Varsani, 2022). The number of CRESS DNA viral genera has increased enormously during the past few years, and the majority of this increase can be attributed to the advances in viral metagenomics (Simmonds et al., 2017). An unforeseen diversity and distribution have been recognized for CRESS DNA viruses, which continue to expand.

The protein content among CRESS DNA viruses differs significantly. While some may have up to ten ORFs, even the most compact genomes have two ORFs: one encoding the Rep and one encoding a CP. All eukaryotic CRESS DNA viruses encode a distinctive homologous Rep that is presumably conserved due to its essential function in viral genome replication through rolling circle replication. In fact, the Rep is often the only gene with homology among the divergent eukaryotic CRESS DNA viruses and thus has been a preferred choice for phylogenetic analyses and higher-level taxonomic classification (Simmonds et al., 2017). The known capsids of CRESS DNA viruses vary in size and shape. Whereas recombination events among CRESS DNA viral CPs are known to be present (Rybicki, 1994), evidence also exists for the shared evolutionary history of CP genes with RNA viruses (Gibbs and Weiller, 1999; Kazlauskas et al., 2017; Krupovic et al., 2009; Roux et al., 2013). CPs are highly divergent compared to the homologous Rep protein shared among the eukaryotic CRESS DNA viruses. While progress has been made in understanding the evolution of eukaryotic CRESS DNA viruses based on the unifying Rep, the evolutionary histories of the CPs seem to be more complex and need to be the subject of future studies. It is highly likely that even more novel species, genera, and families will be discovered in the future, given that these viruses reside in a vast range of hosts and environments and a large part of Earth's virome is still largely unknown.

MECHANISM OF INTEGRATION INTO HOST GENOMES

Viruses can partially or entirely integrate their viral genes into host genomes as endogenous viral elements (EGEs). EGEs are observed in all three domains of life, although they are highly profound in eukaryotes. ssDNA viruses employ different mechanisms to integrate into host genomes and vary between the viruses infecting bacteria, archaea, and eukaryotes.

ssDNA Virus Integration in Bacterial Genomes

Viruses infecting bacteria, called bacteriophages, favor integration into bacterial genomes to propagate their survival, particularly when environmental resources are scarce. Members of the families *Inoviridae* and *Microviridae* are known to integrate into host genomes and are widely recognized. ssDNA viruses employ one of several mechanisms for integration into bacterial genomes:

1) *Serine/tyrosine integrases*: The well-studied serine/threonine integrases are encoded by viruses to initiate a nucleophilic attack on a specific site on the host genome, thereby recombining with them. The recombination mechanisms are significantly different for the two integrases, although they share the same mode of nucleophilic reaction. Tyrosine recombinases cleave and religate one DNA strand from each DNA duplex at a time. Both types of integrases operate on dsDNA molecules. Therefore, the single-stranded viral genome is first converted to a double-stranded replicative intermediate, and the recombination occurs between homologous regions called attachment sites (*att*) on the viral and host genomes.
2) *Host XerC/XerD recombination machinery*: Bacteria utilize the XerC/XerD tyrosine recombinase machinery to resolve their chromosome dimers into monomers by promoting cleavage and ligation at specific sites called the *dif* site. The bacteriophages exploit this machinery to their advantage by recombining between the *att* site on the bacterial genome and the *dif* site on the host genome.
3) *DDE transposases*: Several members of *Inoviridae* are known to encode RNase H-like DDE transposases for integration into the host genome. These encoded transposases have never been experimentally studied, and only hypothetical models have been created to understand their mechanisms of action. The basic mode of action of DDE transposases involves the hydrolysis of the phosphodiester backbone, generating exposed ends, followed by ligation to target DNA.

ssDNA Virus Integration into Archaeal Genomes

Two families, *Pleolipoviridae* and *Spiraviridae*, are known to infect archaea. Although pleolipovirus-related proviruses are known to encode tyrosine recombinase, none of the isolated pleolipoviruses are known to encode integrases. This raises questions on the general propensity of pleolipoviruses to integrate into host genomes. *Spiraviridae*, however, have not been investigated for their integration into the host genome but are known to encode tyrosine recombinase, suggesting an integration into hosts.

ssDNA Virus Integration into Eukaryotic Genomes

Unlike their bacterial and archaeal counterparts, none of the viruses infecting eukaryotes are known to employ similar integration mechanisms. Although widely found integrated into eukaryotes, the mode of integration of several family members is still unknown. Owing to its potential as a vector for gene therapy, adeno-associated virus (AAV) is the only virus that has been extensively studied in this respect.

1) *Parvovirus endonuclease*: AAV integrates into the human genome with a preference for a specific site, termed *AAVS1*, which is located on human chromosome 19 (Smith, 2008). An endonuclease expressed by its Rep gene is the only AAV-encoded protein required for

integration. The Rep creates a nick on a Rep binding site present on the host genome, leading to strand extension. At this stage, the viral genome integrates, proceeding with extension and later the displacement of the new integrated strand. Several parvovirus integration events have been found in the genomes of diverse animals, and therefore, it is likely that the AAV integration paradigm extends to other members of the *Parvoviridae*.

2) *ssDNA circular genome integration*: Several viruses encompassing circular genomes are found to be endogenized in the host genome, but the mechanism of integration remains unknown. The Rep endonucleases are known to mediate the insertion of circular ssDNA into the host, forming transposons as by-products.

PATHOGENIC ssDNA VIRUSES

ssDNA viruses are a considerable threat to all living organisms. Although ssDNA virus families infect hosts across the eukaryotic domain, in the present section we will only discuss the ssDNA viruses that are of human and veterinary importance. Only a few human and animal pathogenic viruses are known that have ssDNA genomes and fall under the *Parvoviridae*, *Circoviridae*, and *Anelloviridae* families. In the last few years, members of the *Smacoviridae* and the newly created *Redondoviridae* were found to be associated with human and animal samples. However, it remains to be established whether these virus families cause overt disease in humans and animals. The emergence of many of the ssDNA viruses as serious pathogens of eukaryotic organisms in recent times can be attributed more to our growing awareness of their presence rather than the appearance of new pathogens for a given host (Rosario et al., 2012).

Parvoviruses are among the smallest and simplest viruses known to date. There are three species among parvoviruses, namely, human parvovirus B19, human bocavirus, and human parvovirus 4, that are known to be pathogenic to humans. The dependoparvoviruses also infect humans but are not associated with any disease and replicate efficiently only in the presence of a helper virus (adenovirus, vaccinia virus, or herpesvirus). This nonpathogenic property of AAVs has been exploited to deliver therapeutic genes to several mammals, including humans. Animal pathogenic parvoviruses exist in much larger numbers and are of great importance in both domestic and livestock husbandry. Some of the animal pathogenic viruses that warrant special mention in this family include canine parvovirus, porcine parvovirus, Aleutian mink disease virus, and feline panleukopenia virus (Table 10.3). Minute virus of mice is a well-studied prototype in the genus *Protoparvovirus* in this family, and most molecular biological processes concerning replication and transcription have been elucidated in this model virus.

The circoviruses and anelloviruses also comprise viruses with single-stranded DNA genomes but have circular DNA genomes, which are linear in the case of parvoviruses. The member viruses of the family *Circoviridae* share some common virion and genome properties but are ecologically, biologically, and antigenically quite distinct. Phylogenetically, members of the *Circoviridae* family are related to viruses from the plant families *Geminiviridae* and *Nanoviridae* (Niagro et al., 1998). Therefore, it is speculated that animal circoviruses may have originated from a plant nanovirus through host-switching and subsequent recombination with a mammalian virus. The family *Circoviridae* is divided into the genera *Circovirus* and *Cyclovirus*. Members of the genus *Circovirus* have been identified in various mammals (chimpanzees, dogs, humans, and pigs), birds, and freshwater fish, and the knowledge of their biology has been gathered mainly from porcine circoviruses. An ever-growing number of species of cycloviruses have been described in mammals (humans, chimpanzees, goats, cattle, bats, pigs, sheep, and camels), birds (chicken and duck), and insects (dragonflies and wood cockroaches). However, in contrast with circoviruses, the pathological significance of cyclovirus infection remains unclear since the discovery of cycloviruses was solely based on the identification of viral DNA by degenerate PCR and metagenomic sequencing (Rosario et al., 2017). Characteristic pathogenic human and animal viruses of the *Circoviridae* family are listed in Table 10.4.

Single-Stranded DNA Viruses

TABLE 10.3

Characteristic Pathogenic Human and Animal Viruses of *Parvoviridae* Family

Subfamily	Genus	Human Virus	Animal Virus
Parvovirinae	*Amdoparvovirus*		Aleutian mink disease virus
			Raccoon dog and fox amdoparvovirus
			Red panda amdoparvovirus
			Skunk amdoparvovirus
	Aveparvovirus		Chicken parvovirus
			Turkey parvovirus
	Bocaparvovirus	Human bocavirus	Minute virus of canines
			Bovine bocaparvoviruses
			Feline bocaviruses 1, 2, 3
			Canine bocavirus 2
	Copiparvovirus		Equine parvovirus-hepatitis
	Dependoparvovirus		Goose parvovirus
			Muscovy duck parvovirus
	Erythroparvovirus	Human parvovirus B19	Simian parvovirus
			Seal parvovirus
	Protoparvovirus		Feline panleukopenia virus
			Porcine parvovirus
			Canine parvovirus
			Rat parvovirus 1
			Minute virus of mice
	Tetraparvovirus	Human parvovirus 4	Porcine parvovirus 2
Hamaparvovirinae	*Chaphamaparvovirus*		Mouse kidney parvovirus
			Tilapia parvovirus

TABLE 10.4

Characteristic Pathogenic Human and Animal Viruses of *Circoviridae* Family

Genus	Human Virus	Animal Virus
Circovirus	Human circovirus 1	Beak and feather disease virus
		Porcine circovirus 2
		Porcine circovirus 3
		Duck circovirus
		Canine circovirus
		Goose circovirus

Torque teno virus (TTV) was the first discovered and most studied anellovirus infecting humans and is currently considered a member of the *Alphatorquevirus* genus. Anelloviruses are also widely distributed in other animal species, such as *Gyrovirus*, which includes chicken anemia virus, a fowl pathogen; *Kappatorquevirus* and *Iotatorquevirus* that infect swine and may cause or exacerbate porcine disease; and *Lambdatorquevirus* that infects sea lions, among others (Table 10.5). Anelloviruses are present worldwide in most healthy people. So far, no illnesses have been definitively associated with their infections. Although first identified in cases of transfusion-transmitted hepatitis, subsequent studies failed to confirm a pathogenic role in hepatitis (Matsumoto et al.,

TABLE 10.5
Characteristic Prototypes of Anelloviruses

Genus	Human Virus	Animal Virus
Alphatorquevirus	Torque teno virus	
Betatorquevirus	Torque teno mini virus	
Epsilontorquevirus		Torque teno tamarin virus
Etatorquevirus		Torque teno felid virus
Gammatorquevirus	Torque teno midi virus	
Gyrovirus		Chicken anemia virus
Iotatorquevirus		Torque teno sus virus 1a
Kappatorquevirus		Torque teno sus virus k2a, k2b
Thetatorquevirus		Torque teno canid virus
Zetatorquevirus		Torque teno douroucouli virus

1999). However, infections may play a role in autoimmunity by changing the homeostatic balance of proinflammatory cytokines and the human immune system, indirectly affecting the severity of diseases caused by other pathogens. Chicken anemia virus, which causes chicken infectious anemia, an immunosuppressive disease mainly characterized by aplastic anemia, growth retardation, lymphoid tissue atrophy, and immunosuppression, leads to huge economic losses to the global poultry industry.

TRANSBOUNDARY MOVEMENT AND THEIR ZOONOTIC POTENTIAL

Globalization and growing demand for livestock and its products have accelerated the movement of live animals and livestock products across borders, which has increased the potential risk of transboundary diseases. Cross-border movement of livestock has facilitated the entry and spread of many diseases in the past. Within the context of ssDNA viruses, the spread of porcine circovirus 2 (PCV2) and beak and feather disease virus (BFDV) are classic examples of transboundary movement. After PCV2 was first discovered in Canadian weaning piglets in 1991, it has been reported in many other countries in Europe, South America, Africa, and South Asia (Afolabi et al., 2017; Karuppannan et al., 2016). Although thought to have originated in Australia, BFDV is now considered to have become distributed worldwide largely due to international legal or illegal trade of captive birds (Fogell et al., 2018). Another ssDNA virus of notable interest is the chicken anaemia virus (CAV). Although initially reported in Japan in 1979, the virus underwent global spreading by intra- and intercontinental migrations and is a poultry pathogen of global concern (Techera et al., 2021). Other ssDNA viruses equally hold importance while regulating the movement of viruses, as transboundary emerging diseases are a major cause of economic losses to livestock rearers. Additionally, these diseases pose a serious threat to the loss of diversity and the extinction of species.

Zoonotic diseases represent a public health problem worldwide, as more than half of the known human pathogens have a zoonotic origin. The potential of the domestic animals–humans–wildlife interface to develop and transmit new infectious agents is a serious consideration and should not be neglected. The ubiquitous presence of ssDNA viruses in diverse habitats suggests that humans are continuously being exposed to these viruses. Many of the most important human pathogens are either zoonotic or originated as zoonoses before adapting to humans. The discoveries of various veterinary-associated circular ssDNA viruses by metagenomics studies of DNA from domestic animals consumed by humans raise a concern about zoonotic transmission (Zhang et al., 2014). Although no convincing biological evidence exists for sustained human infections, PCVs have been identified in cell lines used to prepare vaccines and commercial rotavirus vaccines (Dubin et al.,

Single-Stranded DNA Viruses

2013; Ma et al., 2011). The public health significance of potentially zoonotic porcine viruses seems to be multidimensional, as pig organs are also preferred for xenotransplantation.

CONCLUDING REMARKS

The ssDNA viruses constitute a widespread, diverse, and important group of viruses affecting all three domains of life. An intriguing feature of this group of viruses is their rapid evolutionary potential and minimal protein-coding capacity. ssDNA viruses have gained attention as emerging pathogens in the last few decades and cannot be overlooked anymore. Metagenomic mining and characterization of Rep-like sequences have revolutionized the process and pace of virus discovery, expanding known host ranges and environmental distributions. High mutation rates and recombination, along with the capture of genomic components, have played a significant role in the emergence and evolution of new viral groups, generating diversity within the virosphere. It might not be wrong after all to speculate that ssDNA viruses may have a role in the modulation of global ecology and environment, considering their ubiquitous presence in diverse ecological niches.

However, an enormous task remains in understanding the zoonotic potential of newly discovered ssDNA viruses. There has been ample shreds of evidence of ssDNA virus being associated with human samples. It is just a matter of time before these viruses emerge as serious pathogens for humans. The pace of virus discovery is likely to increase in the near future, and we need to prepare ourselves for the little-known viruses discovered through metagenomics. Although we cannot precisely predict which virus will emerge in humans as a pathogen, we can work together to better understand the viruses that are contenders. Using machine learning models, Mollentze and colleagues (2021) have recently presented an approach for prioritizing novel viruses based on compositional signatures present within viral genomes, which outperformed traditional approaches based on overall measures of relatedness between novel viruses and viruses already known to infect humans. The performance of sequence-based prioritization approaches is likely to improve further as our understanding of the virosphere grows. However, complementary non-genomic triage approaches, like prioritizing viruses commonly encountered at the human–animal interface, are a welcome addition to improving our capacity to characterize zoonotic viruses.

REFERENCES

Afolabi, K.O., Iweriebor, B.C., Okoh, A.I., Obi, L.C., 2017. Global Status of Porcine circovirus Type 2 and Its Associated Diseases in Sub-Saharan Africa. *Adv Virol* 2017, 6807964.

Diemer, G.S., Stedman, K.M., 2012. A novel virus genome discovered in an extreme environment suggests recombination between unrelated groups of RNA and DNA viruses. *Biol Direct* 7, 13.

Dubin, G., Toussaint, J.F., Cassart, J.P., Howe, B., Boyce, D., Friedland, L., Abu-Elyazeed, R., Poncelet, S., Han, H.H., Debrus, S., 2013. Investigation of a regulatory agency enquiry into potential porcine circovirus type 1 contamination of the human rotavirus vaccine, Rotarix: Approach and outcome. *Hum Vaccin Immunother* 9(11), 2398–2408.

Fogell, D.J., Martin, R.O., Bunbury, N., Lawson, B., Sells, J., McKeand, A.M., Tatayah, V., Trung, C.T., Groombridge, J.J., 2018. Trade and conservation implications of new beak and feather disease virus detection in native and introduced parrots. *Conserv Biol* 32(6), 1325–1335.

Gibbs, M.J., Weiller, G.F., 1999. Evidence that a plant virus switched hosts to infect a vertebrate and then recombined with a vertebrate-infecting virus. *Proc Natl Acad Sci USA* 96(14), 8022–8027.

Karuppannan, A.K., Ramesh, A., Reddy, Y.K., Ramesh, S., Mahaprabhu, R., Jaisree, S., Roy, P., Sridhar, R., Pazhanivel, N., Sakthivelan, S.M., Sreekumar, C., Murugan, M., Jaishankar, S., Gopi, H., Purushothaman, V., Kumanan, K., Babu, M., 2016. Emergence of porcine circovirus 2 associated reproductive failure in Southern India. *Transbound Emerg Dis* 63(3), 314–320.

Kazlauskas, D., Dayaram, A., Kraberger, S., Goldstien, S., Varsani, A., Krupovic, M., 2017. Evolutionary history of ssDNA bacilladnaviruses features horizontal acquisition of the capsid gene from ssRNA nodaviruses. *Virology* 504, 114–121.

Kazlauskas, D., Varsani, A., Koonin, E.V., Krupovic, M., 2019. Multiple origins of prokaryotic and eukaryotic single-stranded DNA viruses from bacterial and archaeal plasmids. *Nat Commun* 10(1), 3425.

Koonin, E.V., Krupovic, M., Agol, V.I., 2021. The Baltimore classification of viruses 50 years later: How does it stand in the light of virus evolution? *Microbiol Mol Biol Rev* 85(3), e0005321.

Krupovic, M., Koonin, E.V., 2014. Evolution of eukaryotic single-stranded DNA viruses of the Bidnaviridae family from genes of four other groups of widely different viruses. *Sci Rep* 4, 5347.

Krupovic, M., Ravantti, J.J., Bamford, D.H., 2009. Geminiviruses: a tale of a plasmid becoming a virus. *BMC Evol Biol* 9, 112.

Krupovic, M., Varsani, A., 2022. Naryaviridae, Nenyaviridae, and Vilyaviridae: Three new families of single-stranded DNA viruses in the phylum Cressdnaviricota. *Arch Virol* 167(12), 2907–2921.

Ma, H., Shaheduzzaman, S., Willliams, D.K., Gao, Y., Khan, A.S., 2011. Investigations of porcine circovirus type 1 (PCV1) in vaccine-related and other cell lines. *Vaccine* 29(46), 8429–8437.

Malathi, V.G., Renuka Devi, P., 2019. ssDNA viruses: key players in global virome. *Virusdisease* 30(1), 3–12.

Matsumoto, A., Yeo, A.E., Shih, J.W., Tanaka, E., Kiyosawa, K., Alter, H.J., 1999. Transfusion-associated TT virus infection and its relationship to liver disease. *Hepatology* 30(1), 283–288.

Mollentze, N., Babayan, S.A., Streicker, D.G., 2021. Identifying and prioritizing potential human-infecting viruses from their genome sequences. *PLoS Biol* 19(9), e3001390.

Niagro, F.D., Forsthoefel, A.N., Lawther, R.P., Kamalanathan, L., Ritchie, B.W., Latimer, K.S., Lukert, P.D., 1998. Beak and feather disease virus and porcine circovirus genomes: intermediates between the geminiviruses and plant circoviruses. *Arch Virol* 143(9), 1723–1744.

Rosario, K., Breitbart, M., Harrach, B., Segales, J., Delwart, E., Biagini, P., Varsani, A., 2017. Revisiting the taxonomy of the family Circoviridae: establishment of the genus Cyclovirus and removal of the genus Gyrovirus. *Arch Virol* 162(5), 1447–1463.

Rosario, K., Duffy, S., Breitbart, M., 2012. A field guide to eukaryotic circular single-stranded DNA viruses: Insights gained from metagenomics. *Arch Virol* 157(10), 1851–1871.

Roux, S., Enault, F., Bronner, G., Vaulot, D., Forterre, P., Krupovic, M., 2013. Chimeric viruses blur the borders between the major groups of eukaryotic single-stranded DNA viruses. *Nat Commun* 4, 2700.

Rybicki, E.P., 1994. A phylogenetic and evolutionary justification for three genera of Geminiviridae. *Arch Virol* 139(1–2), 49–77.

Simmonds, P., Adams, M.J., Benko, M., Breitbart, M., Brister, J.R., Carstens, E.B., Davison, A.J., Delwart, E., Gorbalenya, A.E., Harrach, B., Hull, R., King, A.M., Koonin, E.V., Krupovic, M., Kuhn, J.H., Lefkowitz, E.J., Nibert, M.L., Orton, R., Roossinck, M.J., Sabanadzovic, S., Sullivan, M.B., Suttle, C.A., Tesh, R.B., van der Vlugt, R.A., Varsani, A., Zerbini, F.M., 2017. Consensus statement: Virus taxonomy in the age of metagenomics. *Nat Rev Microbiol* 15(3), 161–168.

Smith, R.H., 2008. Adeno-associated virus integration: virus versus vector. *Gene Ther* 15(11), 817–822.

Techera, C., Marandino, A., Tomas, G., Grecco, S., Hernandez, M., Hernandez, D., Panzera, Y., Perez, R., 2021. Origin, spreading and genetic variability of chicken anaemia virus. *Avian Pathol* 50(4), 311–320.

Zhang, W., Li, L., Deng, X., Kapusinszky, B., Delwart, E., 2014. What is for dinner? Viral metagenomics of US store bought beef, pork, and chicken. *Virology* 468–470, 303–310.

Zhao, L., Rosario, K., Breitbart, M., Duffy, S., 2019. Eukaryotic circular rep-encoding single-stranded DNA (CRESS DNA) viruses: Ubiquitous viruses with small genomes and a diverse host range. *Adv Virus Res* 103, 71–133.

11 Double-Stranded RNA Viruses

Nilave Ranjan Bora, Kiran Singh and Sachin Kumar
Department of Biosciences and Bioengineering,
Indian Institute of Technology Guwahati, Guwahati, Assam, India

INTRODUCTION

Double-stranded RNA (dsRNA) viruses have two strands of ribonucleic acid as their genomes. They are polyphyletic and transcribe positive-strand mRNA by the RNA-dependent RNA polymerase (RdRp). dsRNA viruses are divided into the phyla *Duplornaviricota* and *Pisuviricota*. They do not share a common ancestry. According to the Baltimore classification, these are clustered into Group III based on mRNA synthesis [1].

REPLICATION PROPERTIES OF DOUBLE-STRANDED RNA VIRUSES

RNA-dependent RNA polymerase (RdRp), which is virally encoded, is used by RNA viruses to reproduce their genomes. The template for the synthesis of new RNA strands is the RNA genome. Subgenomic mRNAs are also replicated by several RNA viruses. RdRps of viral origin most likely shared an ancestor. The RdRp, along with other proteins, is frequently referred to as the replicase complex as it is involved in the synthesis of viral genomes. The replicase complex may also include RNA helicases (to unwind base-paired portions of the RNA genome) and ATPases (to provide energy for the polymerization process), in addition to the RdRp. Different viral families have a variable number of proteins in replicase complexes. Host proteins also play a role in the replication of the genome.

MAJOR FAMILIES OF dsRNA VIRUSES

dsRNA viruses, according to the International Committee on Taxonomy of Viruses (ICTV), are given in Table 11.1. A taxonomical report of the important family is provided, while an elaborate description of the family is explained later in the chapter.

MEMBERS OF *REOVIRIDAE*

History: The largest family of dsRNA viruses is the *Reoviridae*. It has viruses belonging to 15 genera, each with a genome made up of 9, 10, 11, or 13 linear dsRNA segments. A total of 75 viral species are present in the family, and member viruses have been isolated from a variety of mammals, birds, reptiles, fish, crustaceans, marine protists, insects, ticks, arachnids, plants, and fungi. The diverse genera of reoviruses have amino acid sequence identities of less than 30% according to the amino acid sequences of the RNA-dependent RNA polymerase. Two exceptions exist: *Rotavirus B* only has a 22% amino acid sequence identity with other rotaviruses, and *Aquareovirus* and *Orthoreovirus* have an amino acid sequence identity of up to 42%. The phylogenetic analysis based on polymerase classifies turreted viruses and non-turreted viruses into different clades. However, it is significant that the RdRp's functional domain shares similarities in the family.

Classification and properties: Reoviruses have icosahedral symmetry but can have spherical appearances. The linear dsRNA segments of the viral genome are surrounded by one, two, or three

DOI: 10.1201/9781003369349-11

69

TABLE 11.1
Universal Taxonomy of Double-Stranded (dsRNA) Viruses

Family	Genus	Host
Reoviridae	*Orthoreovirus*	Mammalian orthoreovirus
	Cardoreovirus	Eriocheir sinensis reovirus
	Orbivirus	Bluetongue virus 1
	Rotavirus	Rotavirus A
	Seadornavirus	Banna virus
	Coltivirus	Colorado tick fever virus
	Aquareovirus	Aquareovirus A
Birnaviridae	*Avibirnavirus*	Infectious bursal disease virus
	Aquabirnavirus	Infectious pancreatic necrosis virus

concentric layers of capsid proteins, which have an overall diameter of 60–80 nm. Reoviruses are classified into two subfamilies: *Spinareovirinae* and *Sedoreovirinae*. Virions of the *Spinareovirinae* family have big spikes or turrets located at the 12 icosahedral vertices of the virus or core particle. *Sedoreovirinae* species lack prominent surface projections on their virions or core particles, giving them an almost spherical or "smooth" appearance. Depending on the genus, multiple terms have been used to characterize reovirus particles with various numbers of capsid layers. The transcriptionally active core particle of the spiked viruses (subfamily *Spinareovirinae*) appears to have only a single full capsid layer, to which the projecting spikes or turrets are connected. This capsid layer has been interpreted as having $T=1$ or $T=2$ symmetry. The outer capsid is often formed by an incomplete protein layer with $T=13$ symmetry that surrounds the core and is penetrable by the projections on the core surface [2]. Therefore, these virus particles are typically thought of as having two shells (Figure 11.1).

Genome organization and replication: The genome consists of ten segments of linear dsRNA, which are packaged in equimolar ratios. The segments possess terminal non-translated regions (NTRs) that are shorter at the 5' terminus than at the 3' terminus. The major ORFs vary in length

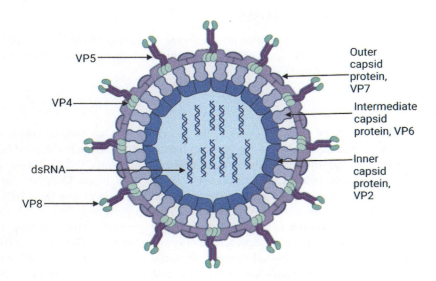

FIGURE 11.1 Typical structure of the rotavirus with important transcription enzyme complex. (Adapted and created from 3, 4.)

from 1059 to 3867 bp. Most of the viral RNA species are monocistronic, although certain segments have additional protein-coding ORFs. Virus entrance mechanisms differ between genera but typically cause the loss of outer capsid components. Transcriptionally active particles are derived from the parental virions in the form of cores (represented by single- or double-layered particles, from the subfamily *Spinareovirinae* or *Sedoreovirinae*, respectively). These active particles are released into the cell cytoplasm. Throughout all stages of infection, these particles repeatedly asymmetrically transcribe full-length mRNA species from each dsRNA segment. The icosahedral edges of these particles serve as the exit points for the mRNA products, which are created in increased copy numbers from the smaller segments. Viral inclusion bodies (VIBs), also known as viroplasms, are discrete structures that appear in the cytoplasm of infected cells. Viroplasms are the sites of viral mRNA synthesis, genome replication, and particle assembly. At the edge of the viroplasms, progeny virus particles appear to be added with outer capsid components, which are anticipated to halt additional mRNA production (Figure 11.2).

The overall course of infection involves adsorption; low pH-dependent penetration and uncoating to core particles; asymmetric transcription of capped, non-polyadenylated mRNAs via a fully conservative mechanism (the nascent strand is displaced); translation; assembly of positive strands into progeny subviral particles; conversion of positive strands to dsRNA; and further rounds of mRNA transcription and translation. The final step of the replication cycle involves the assembly of the outer capsid onto progeny subviral particles to create infectious virions.

The σ1 protein of orthoreovirus mediates attachment to target cells. Virions enter cells via receptor-mediated endocytosis. After virions have been internalized into the cytoplasm of infected cells, they are degraded to core particles, within which virion-associated RNA polymerase (transcriptase) and capping enzymes repetitively transcribe 5′-capped mRNAs that are extruded into the cytoplasm through channels. The negative strands of each dsRNA segment are used as templates by RNA polymerase (transcriptase); initially, only a small subset of genes is transcribed.

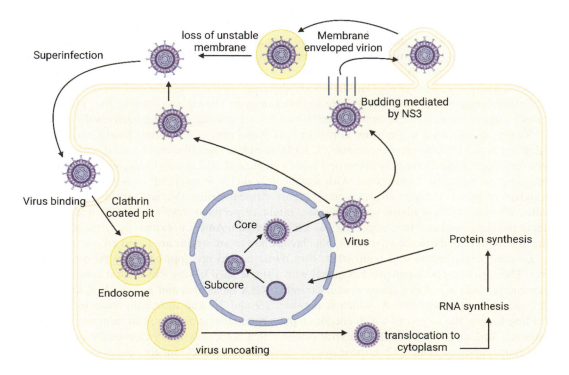

FIGURE 11.2 Schematic representation of reovirus replication cycle. (Adapted and created from 4, 5.)

Upon early mRNA production, nascent offspring subviral particles in the cytoplasm of the infected cells replicate genomic RNA. The mechanism of genomic RNA replication is complex and poorly understood. The newly formed dsRNA then acts as a template for the transcription of additional, uncapped mRNAs. To produce a large pool of viral structural proteins that self-assemble to generate virions, these mRNAs are translated. It is unknown which mechanism allows for the encapsidation of one copy of each dsRNA segment into budding virions. After virus entry, host–cell protein synthesis decreases exponentially.

Orbivirus and rotavirus replication often resemble orthoreovirus replication. Triple-layered viral particles with the outer capsid proteins, VP4 and VP7, are necessary for rotavirus infection. VP4 needs to be proteolytically broken down by chymotrypsin in the small intestine for the virus to enter cells and become more infectious. Although the cellular receptor is unknown, certain rotavirus strains bind to sialic acid residues on the cell surface. Rotavirus can enter cells directly or by receptor-dependent endocytosis. In either case, the act of cell entry removes the virion's outer shell to create the transcriptionally active double-layered virus particle. As rotavirus progenitor particles bud into cisternae of the endoplasmic reticulum of infected cells, they obtain a temporary lipid envelope. This envelope eventually breaks down, leaving VP7 as the primary outer capsid protein.

MEMBERS OF *BIRNAVIRIDAE*

History: The infectious agents of bursal disease in chickens and the pancreatic necrosis of fish are two members of the *Birnaviridae* family that have a substantial economic impact. The first outbreak of infectious bursal disease was identified in 1962 in Gumboro, Delaware; the following outbreaks were then referred to as "Gumboro disease." The cloacal bursa (bursa of Fabricius), where the disease's most prominent lesion can be observed, is where the disease got its current name. During the early stages of the disease's examination, large quantities of virions were seen by electron microscopy in the bursa of affected birds.

In North America, infectious pancreatic necrosis in rainbow trout (*Oncorhynchus mykiss*) was first identified in 1941. However, the viral etiology wasn't established until the 1950s.

Classification and properties: Three genera—*Avibirnavirus*, *Aquabirnavirus*, and *Entomobirnavirus*—make up the family *Birnaviridae*. The sole virus in the genus *Avibirnavirus* is the infectious bursal disease virus. The infectious pancreatic necrosis virus of salmonid fish, associated mollusk, and crustacean viruses are members of the genus *Aquabirnavirus*. Members of *Entomobirnavirus* are exclusively able to infect insects. The classification of the "picobirnaviruses"—that is, viruses that resemble *birnaviruses* but are smaller—has not been resolved [6].

Virions are resistant to exposure at pH 3 to pH 9 and also to ether and chloroform. They are relatively heat stable and can survive at 60°C for 60 minutes. The infectious bursal disease virus has two serotypes (1 and 2), but only serotype 1 is pathogenic and affects only chickens. Three antigenic subgroups of serotype 1 exist, each with significantly different levels of virulence: (1) classic or standard viruses, (2) variant viruses, and (3) highly virulent viruses. Variant viruses cause no mortality, whereas classic or highly virulent viruses might, respectively, result in 10–50% and 50–100% death in young chickens. In Europe, Africa, Asia, and South America, extremely virulent strains of the infectious bursal disease virus exist, but classic and variant strains are found all over the world.

Genome organization and replication: Birnavirus virions are approximately 60 nm in diameter. They are non-enveloped and hexagonal with a single shell having icosahedral symmetry. The genome is made up of two linear double-stranded RNA molecules and is roughly 6 kbp in size (Figure 11.3). One segment, A, which is between 2.9 and 3.4 kbp in length, contains two open reading frames, the largest of which encodes a polyprotein that is processed to form two structural proteins, VP2 and VP3, as well as a viral protease that autocatalytically cleaves the polyprotein. The major capsid protein, VP2, is the major antigenic site responsible for eliciting neutralizing antibodies and is also responsible for cellular tropism and binding. A tiny neutralizing site and group-specific antigenic determinants are present in the inner capsid protein, VP3.

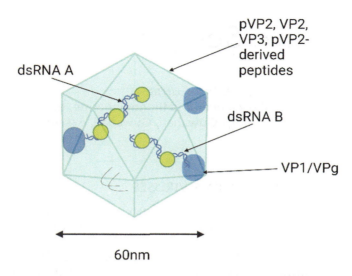

FIGURE 11.3 Schematic diagram of the infectious bursal disease virus (IBDV). (Adapted and created from 4, 5.)

The other segment, B, is approximately 2.7–2.9 kbp in size and encodes VP1, which is the RNA polymerase. A genome-linked protein (VPg) called VP1 circularizes both RNA segments by tightly binding to their ends. Termini of the genome segments resemble those of other segmented RNA viruses in which both the 5′ and 3′ ends are homologous between the segments. Direct terminal and inverted repeats that are expected to create stem and loop secondary structures are present at both ends of both segments. They might also code for signals that are important for replication, transcription, and encapsidation.

Many early events in the infection cycle of *Birnavirus* virions have yet to be characterized. Without significantly hindering cellular RNA or protein synthesis, *birnaviruses* reproduce in the cytoplasm. A virion-associated RNA-dependent RNA polymerase (transcriptase-VP1) is responsible for the transcription of the viral mRNA. RNA replication is expected to be initiated independently at the ends of the segments. The replication is then proceeded by strand displacement, with the inverted terminal repeats at the ends of each segment playing a part in replication.

REFERENCES

1. Koonin, E.V.; Krupovic, M.; Agol, V.I. The Baltimore classification of viruses 50 years later: how does it stand in the light of virus evolution? *Microbiology and Molecular Biology Reviews*, 2021, *85*, e00053–00021.
2. Joklik, W.K. *The Reoviridae*, Springer Science & Business Media, 2013.
3. Flint, J.; Racaniello, V.R.; Rall, G.F.; Hatziioannou, T.; Skalka, A.M. *Principles of virology, Volume 1: Molecular biology*, John Wiley & Sons, 2020.
4. Biorender. *Created with Biorender*, Biorender, 2023.
5. Maclachlan, N.J.; Dubovi, E.J. *Fenner's Veterinary Virology*, Academic Press, 2010.
6. Delmas, B.; Attoui, H.; Ghosh, S.; Malik, Y.S.; Mundt, E.; Vakharia, V.N.; ICTV virus taxonomy profile: Birnaviridae, *Journal of General Virology*, 2019, *100*, 5–6.

12 Positive-Sense RNA Viruses

Kiran Singh and Sachin Kumar
Department of Biosciences and Bioengineering, Indian
Institute of Technology Guwahati, Guwahati, Assam, India

PROPERTIES OF POSITIVE-SENSE RNA VIRUSES

There are eight virus families identified to infect vertebrates and are known to possess single-stranded, positive-sense RNA genomes. These families are further divided based on the presence of an envelope and categorized into the non-enveloped capsid families *Picornaviridae*, *Astroviridae*, *Hepeviridae*, and *Caliciviridae*, as well as the enveloped capsid families *Flaviviridae*, *Togaviridae*, *Arteriviridae*, and *Coronaviridae* (Modrow et al., 2013) (Figure 12.1). Positive-sense RNA viruses use their genome as messenger RNA (mRNA), from which they synthesize one or several polyproteins that are cleaved into individual proteins by viral and cellular proteases. These viruses also possess genetic information for synthesizing RNA-dependent RNA polymerases, which transcribe the positive RNA strand and complementary negative RNA strands, which are the intermediate products of genome replication. This process generates new genomic RNA molecules in the second transcription step. Classification of these viruses into different taxonomic families depends on the number, size, position, and orientation of genes in the RNA; the number of different polyproteins synthesized during viral replication; and the presence of an envelope.

A FEW FAMILIES REPRESENTING POSITIVE-STRANDED RNA GENOMES

PICORNAVIRIDAE

Friedrich Loeffler and Paul Frosch discovered the first picornavirus in 1898, identifying the foot-and-mouth disease (FMD) pathogen (Knowles et al., 2012). In 1909, Karl Landsteiner and Emil Popper published an article reporting the identification of a virus as the pathogen of poliomyelitis, a disease first described by Jacob von Heine in 1840 and later by Oskar Medin. These viruses can cause a cytopathic effect in cultures of human embryonic cells (Enders et al., 1980); however, it was not characterized until 1955 (Von Magnus et al., 1955).

Classification: "Picorna" refers to the family of viruses with a small (pico) RNA genome. Classifying these viruses, known as picornaviruses, into 17 genera is based on their molecular biological properties, sequence homologies, and the diseases they cause (Maclachlan and Dubovi, 2010). Over 150 known human pathogenic viruses belonging to the genus *Enterovirus* have been classified based on new sequence data of their viral genomes. At one time their own genus, human rhinoviruses were recently reclassified within the genus *Enterovirus*, including many different species and types of viruses that can infect animals and humans. This genus is divided into 12 species: enteroviruses A–I and rhinoviruses A, B, and C. All species differ in specific molecular properties, such as the 50 untranslated region (UTR) and the course of infection, from each other. For example, human enteroviruses A and B differ from C and D through distinct sequence elements within the 50 UTR. Poliovirus, similar to human enterovirus C, has been assigned to this species; however, it differs in its ability to cause poliomyelitis from other human enteroviruses. Genome sequencing has revealed that many different types of coxsackieviruses, enteroviruses, and echoviruses have arisen through genetic recombination between various enteroviruses. The animal pathogenic swine vesicular disease virus is a recombinant virus that combines the genomes of human coxsackievirus

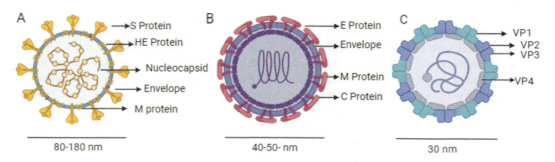

FIGURE 12.1 Representatives of a few essential families of (A) *Coronaviridae*, (B) *Flaviviridae*, and (C) *Picornaviridae* (Biorender, 2023).

B5 and echovirus 9. Aichi virus, a member of the genus *Kobuvirus*, is commonly found in Asia as a pathogen of gastrointestinal infections. Similar symptoms have been observed in cosaviruses and saliviruses. Human rhinoviruses are divided into three species based on differences in the amino acid sequences of their capsid proteins. However, many virus types have not been classified into these species. The genera *Cardiovirus, Sapelovirus, Teschovirus, Erbovirus, Cadicivirus, Aquamavirus*, and *Aphthovirus* primarily comprise animal pathogenic picornaviruses that cause epidemic diseases and have a significant impact on animal breeding and food technology. For example, the FMD virus, classified under the genus *Aphthovirus*, has seven different serotypes worldwide. This genus is named after the blisters or cysts that form in the mucosa of infected animals. The genera *Avihepatovirus, Megrivirus*, and *Tremovirus* have been established as bird isolates. The 50 UTR of this new virus exhibits some similarities to pestiviruses and hepaciviruses. Additionally, the picornaviruses can be categorized into two primary groups based on their molecular properties, which are crucial for their pathogenicity. These groups are (1) acid-stable viruses that can endure the acidic environment of the stomach without losing their infectious properties, infecting the host primarily through the digestive tract, including enteroviruses, parechoviruses, hepatoviruses, avihepatoviruses, sapeloviruses, and teschoviruses; and (2) acid-labile viruses that preferentially infect the host through the upper respiratory tract, such as rhinoviruses, aphthoviruses, and erboviruses.

FLAVIVIRIDAE

Flaviviruses are distinguished by their genome, which is a single-stranded mRNA similar to that of picornaviruses. Similar to picornaviruses, they translate this genome into a single precursor polypeptide comprising both structural and non-structural proteins (Westaway et al., 1985). However, unlike picornaviruses, astroviruses, caliciviruses, and hepeviruses, flaviviruses possess envelopes that surround their capsids and contain viral surface proteins. Comprising several significant pathogens, including the notorious Zika virus, dengue virus, and hepatitis C virus, this family captivates scientists and medical professionals with its complex biology, transmission dynamics, and clinical implications. Flaviviridae encompasses a wide array of viruses, categorized into four genera: *Flavivirus, Pestivirus, Hepacivirus*, and *Pegivirus*. *Flaviviridae* viruses exhibit diverse transmission modes, including vector-borne, parenteral, and sexual transmission. The ecological and environmental factors influencing vector populations play a crucial role in the epidemiology of arthropod-borne flaviviruses, making them particularly challenging to control. Globalization, urbanization, and climate change have contributed to the spread of *Flaviviridae* viruses, increasing their impact on human and animal populations worldwide.

Classification: The family *Flaviviridae* encompasses more than 70 different flavivirus types. The genus *Flavivirus* includes yellow fever virus; the jaundice caused by the virus was eponymous for the denomination of both the family and the genus (*flavus*, Latin for "yellow") (Maclachlan and Dubovi, 2010). It was the first virus for which an insect-associated transmission mode had been seen

76   Textbook of General Virology

(*Aedes* spp.). The *Flavivirus* genus includes several other viruses transmitted by arthropods (insects and arachnids) and pathogenic in humans, such as Zika, dengue, West Nile, and Japanese encephalitis. Arthropod vectors, such as mosquitoes and ticks, primarily transmit these viruses. They are responsible for a range of diseases, from mild febrile illnesses to severe neurological complications and hemorrhagic fever. Pestiviruses mainly infect mammals, with significant implications for veterinary medicine. The most well-known member of this genus is the bovine viral diarrhea virus (BVDV), which causes gastrointestinal and systemic diseases in cattle. Pestiviruses also affect other domestic and wild animals, posing challenges for livestock management and conservation efforts. Hepaciviruses are primarily associated with hepatitis in humans and animals. Hepatitis C virus (HCV), a member of this genus, is a primary global health concern, causing chronic liver diseases. Despite significant advancements in treatment, HCV remains a substantial cause of morbidity and mortality worldwide. Pegiviruses, formerly known as GB viruses, are newly discovered *Flaviviridae* members. These viruses have been found in humans and various non-human primates, and their clinical significance is still under investigation. Pegiviruses are unique in their ability to establish persistent infections without causing significant disease, leading to ongoing research into their potential role in modulating host immune responses.

CORONAVIRIDAE

The *Coronaviridae* family, *Arteriviridae*, and *Roniviridae* are classified under *Nidovirales*. In 1965, David A.J. Tyrrell et al. discovered human coronaviruses during cold epidemics. Initially, these viruses were classified as their own family based on their distinct morphological differences from other viruses, as reported in 1968. Electron micrographs revealed virus particles encircled by an envelope containing embedded proteins, which gave them a "halo" appearance. Later, when the molecular details of their genome structure and replication mechanisms became known, their original classification based on morphological analyses was confirmed (Maclachlan and Dubovi, 2010).

Coronavirus infections typically cause mild upper respiratory tract infections and colds in humans. However, they can also cause acute gastroenteritis in domestic mammals such as cattle, swine, cats, and dogs. Additionally, coronaviruses are associated with other disease patterns, such as encephalitis in pigs and a fatal systemic infection in cats known as feline infectious peritonitis, caused by the feline coronavirus. The mouse hepatitis virus also causes liver inflammation and bronchitis in rodents and is an essential model for understanding pathogenetic mechanisms. In addition to the coronaviruses that affect mammals, certain types can cause severe infections in poultry. One notable example is the avian infectious bronchitis virus. The first occurrences of severe acute respiratory syndrome (SARS)-related coronavirus in humans occurred during the winter of 2002–2003, primarily in Southeast Asian countries such as China, Hong Kong, and Taiwan, as well as Canada. This global epidemic resulted in more than 8,000 reported cases and 700 deaths over several years. The origin of this novel virus was a mystery for a long time until an almost identical SARS-related coronavirus was discovered in China, transmitted by *Rhinolophus* spp. and greater horseshoe bats to civet cats, which can then pass the pathogen to humans in animal markets.

Classification: Coronaviruses are divided into two subfamilies: *Coronavirinae* and *Torovirinae*. Within *Torovirinae*, the white bream virus belongs to the genus *Bafinivirus* and is pathogenic to fish. In contrast, bovine, equine, human, and porcine toroviruses, which are part of the genus *Torovirus*, cause gastrointestinal infections in their respective hosts (Siddell et al., 1983). The biology and diseases associated with toroviruses are not well understood. The subfamily *Coronavirinae* is categorized into four genera based on genome organization and sequence differences. Coronaviruses infecting humans and various mammals, including ungulates, carnivores, and bats, have been placed into the *Alphacoronavirus* and *Betacoronavirus* genera. The SARS-related coronavirus is classified under *Betacoronavirus*, and similar viruses are found in civet cats and bats. The *Gammacoronavirus* and *Deltacoronavirus* genera include species that affect different bird species.

REFERENCES

Biorender, 2023. *Created with Biorender.* Biorender.

Enders, J.F., Robbins, F.C., Weller, T.H., 1980. The cultivation of the poliomyelitis viruses in tissue culture. *Journal of Infectious Diseases*, 2(3), 493–504.

Knowles, N., Hovi, T., Hyypiä, T., King, A., Lindberg, A.M., Pallansch, M., Palmenberg, A., Simmonds, P., Skern, T., Stanway, G., 2012. *Family-Picornaviridae.* The Microbiology Society.

Maclachlan, N.J., Dubovi, E.J., 2010. *Fenner's Veterinary Virology.* Academic Press.

Modrow, S., Falke, D., Truyen, U., Schätzl, H.J., 2013. Viruses with single-stranded, positive-sense RNA genomes. In *Molecular Virology*, p. 185. Springer, Berlin, Heidelberg, https://doi.org/10.1007/978-3-642-20718-1_14.

Siddell, S., Anderson, R., Cavanagh, D., Fujiwara, K., Klenk, H., Macnaughton, M., Pensaert, M., Stohlman, S., Sturman, L., Van Der Zeijst, B.A.M., 1983. Coronaviridae. *Intervirology*, 20(4), 181–189.

Von Magnus, H., Gear, J., Paul, J.R., 1955. A recent definition of poliomyelitis viruses. *Journal of Virology*, 1(2), 185–189.

Westaway, E.G., Brinton, M., Gaidamovich, S.Y., Horzinek, M., Igarashi, A., Kääriäinen, L., Lvov, D., Porterfield, J., Russell, P., Trent, D.J., 1985. Flaviviridae. *Intervirology*, 24(4), 183–192.

13 Minus-Strand RNA Viruses

Kamal Shokeen and Sachin Kumar
Department of Biosciences and Bioengineering, Indian
Institute of Technology Guwahati, Guwahati, Assam, India

INTRODUCTION

As mentioned in the previous chapter, RNA viruses can be classified according to the polarity of the RNA. In addition to plus-strand RNA, numerous families of viruses contain minus-strand RNA as their genome and are grouped into the order *Mononegavirales*. These viruses belong to Group V on the Baltimore classification (Baltimore, 1971). Specifically, seven families of viruses are listed in Table 13.1, of which six constitute a significant threat, being the causative agents of numerous epidemics and pandemics. Most of these viruses cause severe disease and devastating infections in the large population of human and animal hosts but have a low fatality rate. However, viruses like Ebola and rabies confer high fatality rates but within a small fraction of the population. Understanding the infection dynamics in relevant hosts from a One Health perspective is crucial to gaining insights into novel negative-sense zoonotic RNA viruses' emergence, spillover, and disease severity.

They have a unique replication mechanism as their genome, being a single-stranded negative sense, acts as a template to synthesize a complementary positive sense mRNA intermediate. This process is generally carried out by the RNA-dependent RNA polymerase (RdRp) packaged in the virions. These newly synthesized antigenomes are then used to produce viral proteins required to assemble and pack new virus particles (Fields, 2007; Pringle and Easton, 1997; Whelan et al., 2004; Lamb, 2006a).

In this chapter, we shall discuss the general characteristics of negative-sense RNA viruses, their replication cycle, their impact on animal health, and their applications.

GENERAL CHARACTERISTICS

The genomes of these viruses are usually linear and contain inverted terminal repeats, which are palindromic nucleotide sequences at each end (Li et al., 2015). These viruses have a lipid envelope surrounding their nucleocapsid, including the viral genome and associated proteins. The viral particle has a helical structure typically composed of two main components: the viral envelope and nucleocapsid.

Viral envelope: The outermost layer comprises a lipid bilayer surrounding the capsid. It is derived from the host cell membrane, and viral glycoproteins are embedded in it, which enable the virus to attach to host cells and facilitate the fusion of viral and host membranes during entry.

Nucleocapsid: It is a unit of the viral structure consisting of a capsid with an enclosed viral RNA genome and its structural proteins. It safeguards the genome and its transportation into the host cell during entry.

The structural proteins of negative-sense RNA viruses include:

- **Nucleoprotein (NP)**: It is a conjugated protein linked to the viral genome, forming a ribonucleoprotein complex responsible for viral transcription and replication.
- **Polymerase and phosphoprotein (L and P)**: They form a core polymerase primarily responsible for replicating the viral RNA genome and producing new viral RNA molecules.

TABLE 13.1
Families of Negative-Sense RNA Viruses

Family	Genome Size (kb)	Members	Host Range
Arenaviridae	10–14	Lassa virus, Junin virus, Machupo virus, Sabia virus, Guanarito virus, Chapare virus, Whitewater Arroyo virus, Latino virus, Oliveros virus, Tacaribe virus, Bear Canyon virus	Vertebrates
Bornaviridae	9	Borna disease virus	Vertebrates
Bunyaviridae	11–20	Hantavirus, Rift Valley fever virus, Crimean-Congo hemorrhagic fever virus, La Crosse virus, California encephalitis virus, Bunyamwera virus, sandfly fever virus, Heartland virus, Dugbe virus, tomato spotted wilt virus	Vertebrates, arthropods, and plants
Filoviridae	13	Ebola virus, Marburg virus, Lloviu virus, Cuevavirus	Vertebrates
Orthomyxoviridae	10–14.6	Influenza virus (A, B, and C), Thogoto virus, Isavirus	Vertebrates
Paramyxoviridae	16–20	Measles virus, mumps virus, respiratory syncytial virus, Hendra virus, Nipah virus, human parainfluenza viruses, Sendai virus, Newcastle disease virus, Turkey rhinotracheitis virus	Vertebrates
Rhabdoviridae	13–16	Rabies virus, vesicular stomatitis virus, Chandipura virus	Vertebrates, arthropods, and plants

- **Matrix protein (M)**: Many viruses have a layer of protein between their envelope and nucleocapsid, known as matrix protein. It interacts with and helps stabilize the nucleocapsid and is involved in assembling new virus particles. It also inhibits host transcription and shuts down viral RNA synthesis before packaging.
- **Glycoproteins (G)**: These proteins are located on the envelope surface and are responsible for attaching the virus to the host cells and facilitating cell entry. They also play a crucial role in protecting the virus from the host immune system and regulating viral gene expression. They usually project from the envelope as spikes visible in the electron microscope.

BENEFITS OF THE NEGATIVE-SENSE GENOME

- **Control over gene expression**: As these viruses require transcription of the genome into a positive-sense RNA intermediate, they have an additional layer of control over gene expression, which enables them to effectively regulate the level of protein synthesis, thereby optimizing viral replication.
- **Reduced risk of recombination**: It also makes them less susceptible to genetic recombination with host cell RNA, which may produce nonfunctional viral proteins.

One of the critical characteristics of these viruses is that sometimes their genome is divided into segments, with each encoding a different viral protein (Ortin and Martin-Benito, 2015). RNA viruses are more commonly observed to have segmented genomes than DNA viruses. Viruses belonging to the *Bornaviridae*, *Filoviridae*, *Paramyxoviridae*, and *Rhabdoviridae* families have a non-segmented RNA genome with similar organization. Hence, they are grouped into the order of Mononegavirales (*mono* means "single segment" and *nega* means "negative sense"). *Arenaviridae*, *Bunyaviridae*, and *Orthomyxoviridae* families possess segmented genomes ranging from two to

80   Textbook of General Virology

eight parts, respectively. Viruses like influenza package all the segments in one virion, whereas the brome mosaic virus packages them in separate virions (Afonso et al., 2016).

BENEFITS OF THE SEGMENTED GENOME

There are numerous benefits of having a segmented genome, including the following:

- **Reassortment**: Possessing a segmented genome allows for the reassortment or reshuffling of segments between different viruses during replication, leading to a new strain of the virus with novel properties. This possibility of new gene combinations accounts for their rapid evolution. Influenza viruses are known to go through reassortment events, giving rise to new strains capable of causing pandemics. The prime example is the 2009 H1N1 pandemic strain, which is a reassortant of avian, human, and swine influenza viruses (Oshitani, 2009).
- **Genetic diversity**: A segmented genome empowers a single virus population with greater genetic diversity. Multiple variants could arise due to independent mutations that can occur on each of the segments. It ultimately helps the viruses adapt rapidly to changing environments and host immune responses.

Overall, the feature of the segmented genome has played a significant role in viral evolution and spread; however, all the individual genome segments must be packaged into each virus particle, or else the virus will be defective due to loss of genetic information.

FAMILIES OF NEGATIVE-SENSE RNA VIRUSES

Despite having similar types of proteins, the virions exhibit varying degrees of heterogeneity. Five family members often appear spherical under the electron microscope, while some are filamentous (Figure 13.1) (Fermin, 2018). Families of minus-strand RNA viruses are classified based on their genome structure, replication strategies, and host range.

The most well-known types are as follows:

Arenaviruses: The *Arenaviridae* family is classified into four genera: *Antennavirus*, *Hartmanivirus*, *Mammarenavirus*, and *Reptarenavirus*. They are responsible for acute infections, including hemorrhagic fevers, in their natural hosts, rodents, which excrete them through urine and saliva. Some can infect humans as well. They generally exhibit granular or spherical shapes with variable diameters of 40–200 nm and trimeric surface spikes. Two glycoproteins, GP1 and GP2, are embedded in the envelope that surrounds the nucleocapsid. The genome consists of two or three single-stranded RNA segments having an ambisense orientation. The replication mechanism of arenaviruses is poorly understood and is often inferred from the replication cycle of bunyaviruses or influenza viruses. Ribonucleoprotein (RNP) complexes contain antigenomic RNA serving as coding templates for genomic RNA synthesis. Viral proteins are produced from capped and non-polyadenylated mRNAs, where the 5′-cap structure is generally derived by cap-snatching from host cell mRNAs. The virus expresses three to four viral proteins; the most abundant is NP, and the least abundant is L polymerase. Zinc-binding matrix protein is not present in antennaviruses and hartmaniviruses. The glycoproteins (GP1 and GP2) are produced through the posttranslational cleavage of an intracellular precursor, glycoprotein-cell-associated preprotein (GPC), by the cellular S1P/SKI protease. In hartmaniviruses and mammarenaviruses, a third GPC cleavage product, signal peptide, stays attached to the GP complex, resulting in a stable signal peptide (SSP) (Martinez et al., 2007; Glushakova and Lukashevich, 1989).

The infection begins by attaching to the host cell receptor and entering via the endosomal route (Borrow and Oldstone, 1994). Once inside the cell, pH-dependent fusion with late endosomes releases the virion RNP complex into the cytoplasm. The RNP then directs genome replication and

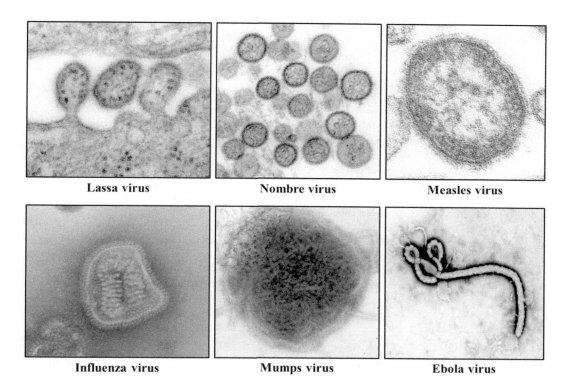

FIGURE 13.1 Transmission electron microscopic images of negative-sense viruses. Images from the Centers for Disease Control and Prevention's Public Health Image Library (PHIL), with identification numbers #8699 (Lassa virus), #1137 (Nombre virus), #8429 (Measles virus), #8420 (Influenza virus), #8758 (Mumps virus), and #1181 (Ebola virus).

viral gene transcription (Meyer et al., 2002). L polymerase reads through the intergenic region (IGR) transcription-terminal signal and produces antigenomic and genomic RNAs (Leung et al., 1977). After the first round of replication, the virus synthesizes subgenomic capped mRNAs that lack terminal poly(A) (Singh et al., 1987). In the case of ambisense orientation, the virus utilizes a cap-snatching mechanism or polymerase slippage to start its transcription using the cellular mRNA's cap. The cap snatching requires an endonuclease to cleave cellular mRNAs to generate a cap leader to prime the transcription step. Virus proteins are then produced from subgenomic capped mRNAs. Last, the virus buds from the cell membrane, which provides an envelope (Garcin and Kolakofsky, 1990; Perez et al., 2003; Gonzalez et al., 2007; Jay et al., 2005; Casals, 1975; Burri et al., 2012; Gunther and Lenz, 2004; Buchmeier et al., 2006).

Filoviruses: The *Filoviridae* family is classified into six genera: *Cuevavirus*, *Dianlovirus*, *Ebolavirus*, *Marburgvirus*, *Striavirus*, and *Thamnovirus*. They are among the most virulent and lethal pathogens responsible for severe hemorrhagic fevers, having a fatality rate of more than 80% in humans. Their genome is 19 kb in size and contains seven genes. The virions have a unique non-uniform, pleomorphous morphology. They are similar to rhabdoviruses but are substantially longer and filamentous. The long filament consists of a helical nucleocapsid having an RNA genome associated with NP, P (VP35), VP30, and L proteins (Table 13.2). The viral particle has a nucleocapsid at its core, surrounded by an envelope with two types of matrix proteins – VP24 (minor) and VP40 (major) – linked to both the inner side of the envelope and the protein components of the nucleocapsid. The envelope also has embedded trimeric glycoproteins, projected approximately 7 nm outside from the surface.

TABLE 13.2
Ebola Virus Proteins and Their Functions

Protein	Molecular Weight (kDa)	Function
GP	120–150	Present on the virus surface as a homotrimer; receptor binding on the host cell; helps in membrane fusion; it is a two-subunit protein; involved in the immune response
L	267	Nucleocapsid component; RNA-dependent RNA polymerase, which helps in transcription
NP	104	Nucleocapsid component; provides integrity to the nucleocapsid complex; associated with the RNA genome
VP24	24	Minor matrix protein associated with envelope and nucleocapsid; inhibits IFN α/β and γ signaling pathway; interacts with STAT1 and inhibits its phosphorylation, which blocks the nuclear transport system
VP30	30	Nucleocapsid component; RNA binding; act as transcription factor; holds RNA template, VP35, and polymerase together; suppresses RNA silencing complex by interacting with DICER1 and TARBP2 proteins
VP35	35	Nucleocapsid component; acts as a cofactor of the polymerase; interacts with NP and controls the free and RNA-bound form of NP; inhibits interferon production
VP40	40	Major matrix protein associated with envelope and nucleocapsid; helps in viral particle packaging and budding; hampers the endosomal trafficking system

The Ebola virus is one of the deadliest viruses known to humans. It causes Ebola virus disease, a severe and often fatal illness characterized by fever, vomiting, diarrhea, and hemorrhaging. The virus is transmitted through direct contact with the bodily fluids of infected individuals (Takada, 2012; Emanuel et al., 2018; Baseler et al., 2017; Banerjee et al., 2021; Feldmann and Klenk, 1996; Peters and Khan, 1999; Sanchez, 2006).

Orthomyxoviruses: The family *Orthomyxoviridae* is classified into six genera: *Influenzavirus A, B, C*; *Isavirus*; *Thogotovirus*; and *Quaranfilvirus*. The genome segments range from six to eight. The influenza viruses are responsible for seasonal flu in humans, a contagious respiratory illness with symptoms including cough, fever, sore throat, runny nose, body aches, headache, and chills. It can also lead to pneumonia in severe cases.

Influenza A is a highly infectious virus affecting many birds and mammals, including humans. It is primarily found in birds, which act as the reservoir. Contrarily, influenza B mainly affects humans, the reservoir for the virus. Influenza C is more diverse than influenza A and B. All A- and B-type influenza viruses' genomes comprise eight RNA segments, amounting to approximately 14 kb, whereas influenza C has only seven RNA segments (Table 13.3). These viruses utilize sialic acid as a receptor in one form or another. Additionally, the enzymes used by the viruses to destroy receptors also differ. All types of influenza viruses can undergo genetic reassortment, thus readily exchanging genetic information (Neumann et al., 2004; Pleschka, 2013; Dowdle et al., 1975; Couch, 1996; Crosby, 2003; Palese, 2006).

Paramyxoviruses: The family *Paramyxoviridae* is classified into two subfamilies: *Paramyxovirinae* and *Pneumovirinae*. They are grouped according to the function of their envelope proteins. *Paramyxovirinae* is organized into seven genera: *Respiroviruses, Rubulaviruses, Avulaviruses, Ferlaviruses, Morbilliviruses, Henipaviruses,* and *Aquaparamyxoviruses*. Their envelope proteins have both neuraminidase and hemagglutination activity, whereas the two genera of *Pneumovirinae* – *Pneumovirus* and *Metapneumovirus* – have neither of these activities. The virions have a diameter of 150–350 nm, with a pleomorphic shape, but usually appear spherical in vitreous ice. They comprise a nucleocapsid surrounded by a lipid envelope containing two transmembrane glycoproteins. They are homo-oligomers and form spike-like projections on the envelope (Table 13.4).

Minus-Strand RNA Viruses

TABLE 13.3
Influenza A Virus Proteins and Their Functions

RNA Segment	Protein	Amino Acid	Function
1	PB2	759	RNA polymerase component; cap recognition
2	PB1	757	RNA polymerase component; elongation
3	PA	716	RNA polymerase component; endonuclease activity; protease
4	HA	566	Surface glycoprotein; hemagglutinin; fusion activity; sialic acid binding; assembly and budding
5	NP	498	Nucleocapsid component; RNA binding and synthesis; RNP nuclear import
6	NA	454	Surface glycoprotein; neuraminidase activity
7	M1	252	Matrix protein; interacts with RNPs and glycoproteins; RNP nuclear export; assembly and budding
8	M2	97	Membrane protein; ion channel activity; assembly and budding
8	NS1	230	Non-structural protein; IFN antagonist activity
8	NS2	121	Non-structural protein; RNP nuclear export; regulation of RNA synthesis

TABLE 13.4
Newcastle Disease Virus Proteins and Their Functions

Protein	Function
N	Nucleocapsid component; provides integrity to the nucleocapsid complex; associated with the RNA genome
P	Virus replication and transcription; helps stabilize L protein; role in virulence
M	Matrix protein; interacts with N protein during viral assembly; responsible for maintaining the spherical shape of the nucleocapsid; viral budding
F	Surface glycoprotein; mediates the fusion to the cell membrane
HN	Surface glycoprotein; receptor recognition; interacts with F to promote fusion; neuraminidase activity
L	Nucleocapsid component; RNA-dependent RNA polymerase, which helps in transcription

They are responsible for various human diseases, including respiratory infections, encephalitis, and measles. The measles virus is a highly contagious virus that causes measles, characterized by fever, cough, runny nose, and a rash that spreads over the entire body. It can also lead to complications such as pneumonia and encephalitis in severe cases. Measles infection starts in the upper respiratory tract but can spread throughout the body, infecting multiple organs. It mainly replicates in lymphoid organs, which reduces T-cell responses resulting in immune suppression. It ultimately increases the risk of secondary infections, which can be severe and life-threatening, and is a significant cause of measles-related mortality (Enders, 1996; Curran and Kolakofsky, 1999; Ganar et al., 2014; Gogoi et al., 2017; Carbone, 2006; Lamb, 2006b).

Rhabdoviruses: The family *Rhabdoviridae* is classified into three subfamilies, 45 genera, and 275 species of viruses with genomes of approximately 10–16 kb in size (Walker et al., 2022). The virions are typically enveloped with bullet-shaped or bacilliform morphology, with particle lengths varying from 130 to 380 nm (Figure 13.2). Their genomes contain only five core genes that encode N, P, M, G, and L proteins but feature an extended 3'-untranslated region following the G mRNA. They are ecologically diverse, with most family members infecting plants or animals, including mammals, birds, reptiles, or fish. Two genera, *Vesiculovirus* and *Lyssavirus*, have been extensively

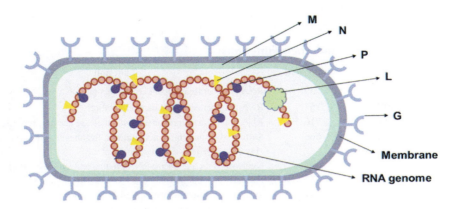

FIGURE 13.2 Schematic diagram of rabies virus particle.

studied, which include viruses like vesicular stomatitis virus (VSV) and rabies virus (RABV). There are currently seven known genotypes or species of *Lyssavirus*.

The first genotype, the classical rabies virus, causes acute encephalomyelitis with a high fatality rate. The RABV genome is relatively simple, consisting of a 58-nucleotide leader sequence, followed by five genes in the order 3′-N-P-M-G-L-5′, separated by non-transcribed intergenic regions and followed by a 57–70 nucleotide trailer. It is a neurotropic pathogen commonly transmitted through the saliva of infected animals, usually via a bite or wound contamination (Fooks et al., 2017). Infection begins in tissues around the bite site, and in due course, it reaches the motor or sensory neurons and subsequently the central nervous system by following neuronal connections. The asymptomatic incubation period can differ and typically lasts around two months, whereas the symptomatic clinical period is brief and intense, lasting approximately one week (Udow et al., 2013). Natural rabies is characterized by severe neurological signs and fatal outcomes with relatively mild neuropathologic changes in the CNS. The development of vaccines aims to safeguard mainly humans and domesticated dogs at risk of contracting the rabies virus. Oral rabies vaccination (ORV) has been used for many years to control the spread of the disease. These ORV baits contain a sachet or plastic packet of the Raboral V-RG rabies vaccine, which includes an attenuated ("modified-live") recombinant vaccinia virus vector vaccine expressing the rabies virus glycoprotein gene (V-RG) (Jackson, 2002, 2003; Lyles, 2006; Nishizono and Yamada, 2012; Maki et al., 2017).

GENOME REPLICATION AND TRANSCRIPTION

The life cycle of minus-strand RNA viruses varies from that of other RNA viruses in numerous ways. They replicate in the host cell using a unique mechanism involving the production of a complementary plus-strand RNA intermediate (Figure 13.3). Their naked genomes are not infectious; they must also deliver the RNA-dependent RNA polymerase into the host to initiate mRNA synthesis. The replication cycle begins with the attachment of the virus to the host cell, facilitated by viral glycoproteins in an envelope that binds to specific host cell surface receptors. The host cell receptors for some viruses are listed in Table 13.5. The binding prompts receptor-mediated endocytosis and leads to nucleocapsid entry into the cytoplasm.

After that, the viral RNA genome is released, and RdRp initiates replication by binding to a 3′ end leader sequence in the genome and produces a positive-sense antigenome. It then serves as a template for synthesizing new genomic negative-sense RNA. In viruses with segmented genomes, replication occurs in the nucleus, and the RdRp produces one mono-cistronic mRNA strand from each genome segment. In viruses with non-segmented genomes, multiple mono-cistronic mRNAs arise using a single-entry site for RdRp at the 3′ end of the genome in the cytoplasm. RdRp

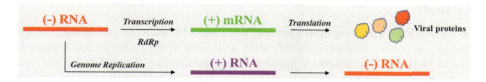

FIGURE 13.3 Schematic diagram of genome replication and protein production.

TABLE 13.5
List of Viruses with Their Host Cell Receptors

Virus	Host Cell Receptor	References
Ebola virus	T-cell immunoglobulin mucin 1 (TIM-1), Niemann-Pick C1 (NPC1), and toll-like receptor 4 (TLR4)	(Kondratowicz et al., 2011; Carette et al., 2011; Okumura et al., 2010)
Hantavirus	β3 integrin, decay-accelerating factor (DAF), and N-linked glycosylation	(Rafikov, 1975; Buranda et al., 2010)
Hendra virus	Ephrin-B2 and ephrin-B3	(Bonaparte et al., 2005; Xu et al., 2012)
Influenza virus	Sialic acid (varies by subtype and species)	(Haff and Stewart, 1965; Springer et al., 1969)
Junin virus	Transferrin receptor 1	(Radoshitzky et al., 2007)
Lassa fever virus	α-Dystroglycan	(Cao et al., 1998)
Machupo virus	Transferrin receptor 1	(Radoshitzky et al., 2007)
Marburg virus	T-cell immunoglobulin mucin 1 (TIM-1), Niemann-Pick C1 (NPC1), and toll-like receptor 4 (TLR4)	(Kondratowicz et al., 2011; Carette et al., 2011; Okumura et al., 2010)
Measles virus	CD46, signaling lymphocytic activation molecule (SLAM), and nectin-4	(Yanagi, 1995; Dhiman et al., 2004; Noyce and Richardson, 2012)
Newcastle disease virus	Sialic acid	(Tsvetkova and Lipkind, 1973)
Nipah virus	Ephrin-B2 and ephrin-B3	(Bonaparte et al., 2005; Xu et al., 2012)
Rabies virus	Acetylcholine receptor and neural cell adhesion molecule (NCAM)	(Thoulouze et al., 1998)

---

recognizes conserved start (Gene Start) and stop (Gene End) signals located at the beginning and end of each gene, respectively, to create discrete mRNAs. Its location with the single-entry site regulates the mRNA levels synthesized for a particular viral gene. The mRNA synthesis is performed sequentially, and attenuation occurs at each gene junction. Therefore, many mRNAs are generated for the genes located toward the 3′ ends of the genome, resulting in high protein production. The N protein, which encapsulates the genome to form the ribonucleoprotein complex, is produced in considerable quantities, while the RNA polymerase is made in minor amounts. The replication of the genomic RNA and the synthesis of mRNAs take place in nucleocapsids (RNPs) containing phosphoprotein and large polymerase (Luo et al., 2020). Replication can only progress through ongoing viral protein synthesis to produce the new proteins required to encapsulate the genome or antigenome.

Most of the mRNAs from viral genes are translated into a single protein. However, a few genes (mainly the P gene) produce mRNAs that can be translated into multiple products using alternative translation initiation codons or alternative splicing. The translation strategy employed by negative-sense RNA viruses differs from that of positive-sense RNA viruses in that they do not generate polyproteins that require further processing. However, in some cases, the glycoproteins are produced as precursors that are processed by cellular enzymes, and some of these can be considered polyproteins.

APPLICATIONS

Negative-sense RNA viruses are used extensively in research, medicine, and biotechnology (Table 13.6). The advancements in genetic modification and engineering techniques for negative-strand RNA viruses have yielded valuable information in understanding the molecular mechanisms in viral infectivity, virus–host interactions, viral pathogenesis, and developing new vaccine candidates and gene therapies. These viruses are ideal targets for reverse genetics as their genomes comprise a single RNA strand that can be easily manipulated and transcribed in vitro.

Recombinant viruses have been exploited as vectors for devising targeted therapies for non-viral diseases such as malignancies and gene therapy for inherited disorders. Developing a recombinant virus is a powerful tool that can be exploited to directly influence a viral isolate's virulence. In addition, reverse genetics of negative-sense RNA viruses can be used to engineer the viral genome to stably express foreign genes and can also be used as a vaccine vector (Krishnamurthy et al., 2000). Recombinant NDV has been used successfully as a vaccine vector for animals (Debnath et al., 2020; Murr et al., 2020) and human pathogens (Sun et al., 2020; Yoshida et al., 2019; Kim et al., 2022). This technique involves constructing a plasmid vector containing a cDNA copy of the viral genome under the control of a suitable promoter. The plasmid can be transfected into cells to produce an utterly infectious virus genetically identical to the original. The infectious clone system has been used to study the replication and pathogenesis of several negative-sense RNA viruses, including influenza, Ebola, and rabies. Rhabdoviruses and measles viruses have been employed as expression vectors for treating various diseases, including cancers, hemophilia, and infectious diseases (Edenborough and Marsh, 2014).

TABLE 13.6
Application of Negative-Sense RNA Viruses with Examples

Genetic Engineering Technique	Virus	Description	Examples	References
Attenuation	Measles virus	Reducing the virulence or pathogenicity to create a live attenuated virus vaccine	Live attenuated measles virus vaccines	(Griffin, 2018)
Biodefense	Ebola and Marburg viruses	Developing treatments and diagnostic tools for diseases caused by these viruses	Ebola virus	(Rasmussen, 2018; Bramwell et al., 2005)
Chimeric virus construction	Ebola virus	Generating novel viruses with desirable characteristics by combining parts from different viruses	Chimeric Ebola virus vaccines	(Bukreyev et al., 2009)
Deletion mutagenesis	Rabies virus	Removing a specific sequence or gene from the genome to create a modified virus	Modified live rabies virus vaccines	(Chatterjee et al., 2018; Sato et al., 2015)
Gene therapy	NDV	Delivering therapeutic genes to cells, which can help treat a variety of diseases	rNDV targeting prostate-specific antigens	(Huang et al., 2020),
Recombinant protein expression	Hantavirus, NDV	Incorporating a gene encoding a foreign protein into host cells to produce the protein	rNDV-Viperin	(Shah et al., 2019)
Vaccine development	NDV	Expressing the antigens of interest can stimulate an immune response without causing disease	rNDV-Ejev, rNDV-NS1jev, rNDV-CSFV, E2 and E^{rns}	(Nath et al., 2020, Kumar et al., 2019)

CONCLUSION

Negative-sense RNA viruses are a diverse group responsible for many significant human and animal diseases. These viruses have a unique genome structure and replication cycle that require an RNA-dependent RNA polymerase to synthesize a positive-sense RNA intermediate to produce viral proteins. Despite the development of vaccines and antiviral therapies, negative-sense RNA viruses have significantly threatened human health in recent years. Further research is needed to better understand the biology of these viruses and develop more effective treatments and preventative measures. On top of that, they have been extensively used for the vaccine development of numerous other viruses and in gene therapy. However, as our understanding of these viruses continues to improve, we can expect to see even more innovative applications of negative-sense RNA genome viruses in the future.

REFERENCES

Afonso, C. L., Amarasinghe, G. K., Bányai, K., Bào, Y., Basler, C. F., Bavari, S., Bejerman, N., Blasdell, K. R., Briand, F.-X. & Briese, T. J. A. O. V. 2016. Taxonomy of the order mononegavirales: update 2016. 161, 2351–2360.

Baltimore, D. 1971. Expression of animal virus genomes. *Bacteriol Rev,* 35, 235–41.

Banerjee, G., Shokeen, K., Chakraborty, N., Agarwal, S., Mitra, A., Kumar, S. & Banerjee, P. 2021. Modulation of immune response in Ebola virus disease. *Curr Opin Pharmacol,* 60, 158–167.

Baseler, L., Chertow, D. S., Johnson, K. M., Feldmann, H. & Morens, D. M. 2017. The pathogenesis of ebola virus disease. *Annu Rev Pathol,* 12, 387–418.

Bonaparte, M. I., Dimitrov, A. S., Bossart, K. N., Crameri, G., Mungall, B. A., Bishop, K. A., Choudhry, V., Dimitrov, D. S., Wang, L. F., Eaton, B. T. & Broder, C. C. 2005. Ephrin-B2 ligand is a functional receptor for Hendra virus and Nipah virus. *Proc Natl Acad Sci U S A,* 102, 10652–7.

Borrow, P. & Oldstone, M. B. 1994. Mechanism of lymphocytic choriomeningitis virus entry into cells. *Virology,* 198, 1–9.

Bramwell, V. W., Eyles, J. E. & Oya Alpar, H. 2005. Particulate delivery systems for biodefense subunit vaccines. *Adv Drug Deliv Rev,* 57, 1247–65.

Buchmeier M. J., D. L. T., J.-C. & Peters, C. J. 2006. Arenaviridae: the viruses and their replication. In: D. M. Knipe & P. M. Howley (eds.) *Fields Virology,* 5th ed. Philadelphia: Lippincott Williams, Wilkins.

Bukreyev, A., Marzi, A., Feldmann, F., Zhang, L., Yang, L., Ward, J. M., Dorward, D. W., Pickles, R. J., Murphy, B. R., Feldmann, H. & Collins, P. L. 2009. Chimeric human parainfluenza virus bearing the Ebola virus glycoprotein as the sole surface protein is immunogenic and highly protective against Ebola virus challenge. *Virology,* 383, 348–61.

Buranda, T., Wu, Y., Perez, D., Jett, S. D., Bonduhawkins, V., Ye, C., Edwards, B., Hall, P., Larson, R. S., Lopez, G. P., Sklar, L. A. & Hjelle, B. 2010. Recognition of decay accelerating factor and alpha(v) beta(3) by inactivated hantaviruses: toward the development of high-throughput screening flow cytometry assays. *Anal Biochem,* 402, 151–60.

Burri, D. J., Da Palma, J. R., Kunz, S. & Pasquato, A. 2012. Envelope glycoprotein of arenaviruses. *Viruses,* 4, 2162–81.

Cao, W., Henry, M. D., Borrow, P., Yamada, H., Elder, J. H., Ravkov, E. V., Nichol, S. T., Compans, R. W., Campbell, K. P. & Oldstone, M. B. 1998. Identification of alpha-dystroglycan as a receptor for lymphocytic choriomeningitis virus and Lassa fever virus. *Science,* 282, 2079–81.

Carbone, K. M. & Rubin, S. A 2006. Mumps virus. In: D. M. Knipe & P. M. Howley (eds.) *Fields Virology,* 5th ed. Philadelphia: Lippincott Williams, Wilkins.

Carette, J. E., Raaben, M., Wong, A. C., Herbert, A. S., Obernosterer, G., Mulherkar, N., Kuehne, A. I., Kranzusch, P. J., Griffin, A. M., Ruthel, G., Dal Cin, P., Dye, J. M., Whelan, S. P., Chandran, K. & Brummelkamp, T. R. 2011. Ebola virus entry requires the cholesterol transporter Niemann-Pick C1. *Nature,* 477, 340–3.

Casals, J. 1975. Arenaviruses. *Yale J Biol Med,* 48, 115–40.

Chatterjee, S., Sullivan, H. A., Maclennan, B. J., Xu, R., Hou, Y., Lavin, T. K., LEA, N. E., Michalski, J. E., Babcock, K. R., Dietrich, S., Matthews, G. A., Beyeler, A., Calhoon, G. G., Glober, G., Whitesell, J. D., Yao, S., Cetin, A., Harris, J. A., Zeng, H., Tye, K. M., Reid, R. C. & Wickersham, I. R. 2018. Nontoxic, double-deletion-mutant rabies viral vectors for retrograde targeting of projection neurons. *Nat Neurosci,* 21, 638–46.

Couch, R. B. 1996. Orthomyxoviruses. In: S. Baron (ed.) *Medical Nicrobiology*, 4th ed. New York: Churchill Livingstone.

Crosby, A. W. 2003. *America's Forgotten Pandemic: The Influenza of 1918*. Cambridge: Cambridge University Press.

Curran, J. & Kolakofsky, D. 1999. Replication of paramyxoviruses. *Adv Virus Res*, 54, 403–22.

Debnath, A., Pathak, D. C., D'silva A, L., Batheja, R., Ramamurthy, N., Vakharia, V. N., Chellappa, M. M. & Dey, S. 2020. Newcastle disease virus vectored rabies vaccine induces strong humoral and cell mediated immune responses in mice. *Vet Microbiol*, 251, 108890.

Dhiman, N., Jacobson, R. M. & Poland, G. A. 2004. Measles virus receptors: SLAM and CD46. *Rev Med Virol*, 14, 217–29.

Dowdle, W. R., Davenport, F. M., Fukumi, H., Schild, G. C., Tumova, B., Webster, R. G. & Zakstelskaja, L. Y. 1975. Orthomyxoviridae. *Intervirology*, 5, 245–51.

Edenborough, K. & Marsh, G. A. 2014. Reverse genetics: unlocking the secrets of negative sense RNA viral pathogens. *World J Clin Infect Dis*, 4, 16–26.

Emanuel, J., Marzi, A. & Feldmann, H. 2018. Filoviruses: ecology, molecular biology, and evolution. *Adv Virus Res*, 100, 189–221.

Enders, G. 1996. Paramyxoviruses. In: S. Baron (ed.) *Medical Microbiology*. 4th ed. Galveston, TX.

Feldmann, H. & Klenk, H. D. 1996. Marburg and Ebola viruses. *Adv Virus Res*, 47, 1–52.

Fermin, G. J. V. 2018. Virion structure, genome organization, and taxonomy of viruses. In *Viruses*, 17–54. doi: 10.1016/B978-0-12-811257-1.00002-4. Epub 2018 March 30. PMCID: PMC7149880.

Fields, B. N. 2007. *Fields' Virology*. Samoa: Wolters Kluwer Health/Lippincott Williams & Wilkins, Philadelphia.

Fooks, A. R., Cliquet, F., Finke, S., Freuling, C., Hemachudha, T., Mani, R. S., Muller, T., Nadin-Davis, S., Picard-Meyer, E., Wilde, H. & Banyard, A. C. 2017. Rabies. *Nat Rev Dis Primers*, 3, 17091.

Ganar, K., Das, M., Sinha, S. & Kumar, S. 2014. Newcastle disease virus: current status and our understanding. *Virus Res*, 184, 71–81.

Garcin, D. & Kolakofsky, D. 1990. A novel mechanism for the initiation of Tacaribe arenavirus genome replication. *J Virol*, 64, 6196–203.

Glushakova, S. E. & Lukashevich, I. S. 1989. Early events in arenavirus replication are sensitive to lysosomotropic compounds. *Arch Virol*, 104, 157–61.

Gogoi, P., Ganar, K. & Kumar, S. 2017. Avian paramyxovirus: a brief review. *Transbound Emerg Dis*, 64, 53–67.

Gonzalez, J. P., Emonet, S., De Lamballerie, X. & Charrel, R. 2007. Arenaviruses. *Curr Top Microbiol Immunol*, 315, 253–88.

Griffin, D. E. 2018. Measles vaccine. *Viral Immunol*, 31, 86–95.

Gunther, S. & Lenz, O. 2004. Lassa virus. *Crit Rev Clin Lab Sci*, 41, 339–90.

Haff, R. F. & Stewart, R. C. 1965. Role of sialic acid receptors in adsorption of influenza virus to chick embryo cells. *J Immunol*, 94, 842–51.

Huang, Z., Liu, M. & Huang, Y. 2020. Oncolytic therapy and gene therapy for cancer: recent advances in antitumor effects of Newcastle disease virus. *Discov Med*, 30, 39–48.

Jackson, A. C. 2002. Rabies pathogenesis. *J Neurovirol*, 8, 267–269.

Jackson, A. C. J. J. O. N. 2003. Rabies virus infection: an update. *J Neurovirol*, 9, 253–258.

Jay, M. T., Glaser, C. & Fulhorst, C. F. 2005. The arenaviruses. *J Am Vet Med Assoc*, 227, 904–15.

Kim, S. H., Shirvani, E. & Samal, S. 2022. Avian paramyxoviruses as vectors for vaccine development. *Methods Mol Biol*, 2411, 63–73.

Kondratowicz, A. S., Lennemann, N. J., Sinn, P. L., Davey, R. A., Hunt, C. L., Moller-Tank, S., Meyerholz, D. K., Rennert, P., Mullins, R. F., Brindley, M., Sandersfeld, L. M., Quinn, K., Weller, M., Mccray, P. B., JR., Chiorini, J. & Maury, W. 2011. T-cell immunoglobulin and mucin domain 1 (TIM-1) is a receptor for Zaire ebolavirus and lake victoria marburgvirus. *Proc Natl Acad Sci U S A*, 108, 8426–31.

Krishnamurthy, S., Huang, Z. & Samal, S. K. 2000. Recovery of a virulent strain of newcastle disease virus from cloned cDNA: expression of a foreign gene results in growth retardation and attenuation. *Virology*, 278, 168–82.

Kumar, R., Kumar, V., Kekungu, P., Barman, N. N. & Kumar, S. 2019. Evaluation of surface glycoproteins of classical swine fever virus as immunogens and reagents for serological diagnosis of infections in pigs: a recombinant Newcastle disease virus approach. *Arch Virol*, 164, 3007–3017.

Lamb, R. A. 2006a. Mononegavirales. In: M. K. Howley (ed.) *Fields Virology*, 5th ed. Lippincott Williams & Wilkins.

Lamb, R. A., and Parks, G. D. 2006b. Paramyxoviridae: the viruses and their replication. In: M. K. Howley (ed.) *Fields Virology*, 5th ed. Wolters Kluwer Health/Lippincott Williams & Wilkins, Philadelphia.

Leung, W. C., Ghosh, H. P. & Rawls, W. E. 1977. Strandedness of Pichinde virus RNA. *J Virol*, 22, 235–7.

Li, C. X., Shi, M., Tian, J. H., Lin, X. D., Kang, Y. J., Chen, L. J., Qin, X. C., Xu, J., Holmes, E. C. & Zhang, Y. Z. 2015. Unprecedented genomic diversity of RNA viruses in arthropods reveals the ancestry of negative-sense RNA viruses. *Elife*, 4, January 29. e05378. doi: 10.7554/eLife.05378. PMID: 25633976; PMCID: PMC4384744..

Luo, M., Terrell, J. R. & Mcmanus, S. A. 2020. Nucleocapsid structure of negative strand RNA virus. *Viruses*, July 30, 12(8), 835. doi: 10.3390/v12080835. PMID: 32751700; PMCID: PMC7472042.

Lyles, D. S. & Rupprecht, C. E. 2006. Rhabdoviridae. In: D. M. K. A. P. M. Howley (ed.) *Fields Virology*, 5th ed. Wolters Kluwer Health/Lippincott Williams & Wilkins, Philadelphia.

Maki, J., Guiot, A. L., Aubert, M., Brochier, B., Cliquet, F., Hanlon, C. A., King, R., Oertli, E. H., Rupprecht, C. E., Schumacher, C., Slate, D., Yakobson, B., Wohlers, A. & Lankau, E. W. 2017. Oral vaccination of wildlife using a vaccinia-rabies-glycoprotein recombinant virus vaccine (RABORAL V-RG((R))): a global review. *Vet Res*, 48, 57.

Martinez, M. G., Cordo, S. M. & Candurra, N. A. 2007. Characterization of Junin arenavirus cell entry. *J Gen Virol*, 88, 1776–1784.

Meyer, B. J., De La Torre, J. C. & Southern, P. J. 2002. Arenaviruses: genomic RNAs, transcription, and replication. *Curr Top Microbiol Immunol*, 262, 139–57.

Murr, M., Hoffmann, B., Grund, C., Romer-Oberdorfer, A. & Mettenleiter, T. C. 2020. A novel recombinant newcastle disease virus vectored DIVA vaccine against peste des petits ruminants in goats. *Vaccines (Basel)*, April 28, 8(2), 205. doi: 10.3390/vaccines8020205. PMID: 32354145; PMCID: PMC7348985.

Nath, B., Vandna, Saini, H. M., Prasad, M. & Kumar, S. 2020. Evaluation of Japanese encephalitis virus E and NS1 proteins immunogenicity using a recombinant Newcastle disease virus in mice. *Vaccine*, 38, 1860–1868.

Neumann, G., Brownlee, G. G., Fodor, E. & Kawaoka, Y. 2004. Orthomyxovirus replication, transcription, and polyadenylation. *Curr Top Microbiol Immunol*, 283, 121–43.

Nishizono, A. & Yamada, K. 2012. Rhabdoviruses. *Uirusu*, 62, 183–96.

Noyce, R. S. & Richardson, C. D. 2012. Nectin 4 is the epithelial cell receptor for measles virus. *Trends Microbiol*, 20, 429–39.

Okumura, A., Pitha, P. M., Yoshimura, A. & Harty, R. N. 2010. Interaction between Ebola virus glycoprotein and host toll-like receptor 4 leads to induction of proinflammatory cytokines and SOCS1. *J Virol*, 84, 27–33.

Ortin, J. & Martin-Benito, J. 2015. The RNA synthesis machinery of negative-stranded RNA viruses. *Virology*, 479–480, 532–44.

Oshitani, H. 2009. Influenza pandemic (H1N1) 2009. *Uirusu*, 59, 139–44.

Palese, P. & Shaw, M. L. 2006. Orthomyxoviridae: The viruses and their replication. In: M. K. Howley (ed.) *Fields Virology*, 5th ed. Wolters Kluwer Health/Lippincott Williams & Wilkins, Philadelphia.

Perez, M., Craven, R. C. & De La Torre, J. C. 2003. The small RING finger protein Z drives arenavirus budding: implications for antiviral strategies. *Proc Natl Acad Sci U S A*, 100, 12978–83.

Peters, C. J. & Khan, A. S. 1999. Filovirus diseases. *Curr Top Microbiol Immunol*, 235, 85–95.

Pleschka, S. 2013. Overview of influenza viruses. *Curr Top Microbiol Immunol*, 370, 1–20.

Pringle, C. R. & Easton, A. J. 1997. *Monopartite Negative Strand RNA Genomes. Seminars in Virology*. Elsevier BV, 49–57.

Radoshitzky, S. R., Abraham, J., Spiropoulou, C. F., Kuhn, J. H., Nguyen, D., LI, W., Nagel, J., Schmidt, P. J., Nunberg, J. H., Andrews, N. C., Farzan, M. & Choe, H. 2007. Transferrin receptor 1 is a cellular receptor for New World haemorrhagic fever arenaviruses. *Nature*, 446, 92–6.

Rafikov, A. M. 1975. Indirect assessment of the state of cerebral circulation by the method of superficial temporal artery oscillography. *Zh Nevropatol Psikhiatr Im S S Korsakova*, 75, 1167–70.

Rasmussen, A. L. 2018. Host factors involved in ebola virus replication. *Curr Top Microbiol Immunol*, 419, 113–150.

Sanchez, A., Geisbert, T. W. & Feldmann, H. 2006. Filoviridae: marburg and Ebola viruses. In: M. K. Howley (ed.) *Fields Virology*, 5th ed..Lippincott Williams & Wilkins, Philadelphia.

Sato, S., Ohara, S., Tsutsui, K. & Iijima, T. 2015. Effects of G-gene deletion and replacement on rabies virus vector gene expression. *PLoS One*, 10, e0128020.

Shah, M., Bharadwaj, M. S. K., Gupta, A., Kumar, R. & Kumar, S. 2019. Chicken viperin inhibits Newcastle disease virus infection in vitro: a possible interaction with the viral matrix protein. *Cytokine*, 120, 28–40.

Singh, M. K., Fuller-Pace, F. V., Buchmeier, M. J. & Southern, P. J. 1987. Analysis of the genomic L RNA segment from lymphocytic choriomeningitis virus. *Virology,* 161, 448–56.

Springer, G. F., Schwick, H. G. & Fletcher, M. A. 1969. The relationship of the influenza virus inhibitory activity of glycoproteins to their molecular size and sialic acid content. *Proc Natl Acad Sci U S A,* 64, 634–41.

Sun, W., Leist, S. R., Mccroskery, S., Liu, Y., Slamanig, S., Oliva, J., Amanat, F., Schafer, A., Dinnon, K. H., 3RD, Garcia-Sastre, A., Krammer, F., Baric, R. S. & Palese, P. 2020. Newcastle disease virus (NDV) expressing the spike protein of SARS-CoV-2 as a live virus vaccine candidate. *EBioMedicine,* 62, 103132.

Takada, A. 2012. Filoviruses. *Uirusu,* 62, 197–208.

Thoulouze, M. I., Lafage, M., Schachner, M., Hartmann, U., Cremer, H. & Lafon, M. 1998. The neural cell adhesion molecule is a receptor for rabies virus. *J Virol,* 72, 7181–90.

Tsvetkova, I. V. & Lipkind, M. A. 1973. Studies on the role of myxovirus neuraminidase in virus-cell receptor interaction by means of direct determination of sialic acid split from cells. 3. One-step growth kinetics of accumulation of the sialic acid liberated from NDV-infected chick embryo cells. *Arch Gesamte Virusforsch,* 42, 125–38.

Udow, S. J., Marrie, R. A. & Jackson, A. C. 2013. Clinical features of dog- and bat-acquired rabies in humans. *Clin Infect Dis,* 57, 689–96.

Walker, P. J., Freitas-Astua, J., Bejerman, N., Blasdell, K. R., Breyta, R., Dietzgen, R. G., Fooks, A. R., Kondo, H., Kurath, G., Kuzmin, I. V., Ramos-Gonzalez, P. L., Shi, M., Stone, D. M., Tesh, R. B., Tordo, N., Vasilakis, N., Whitfield, A. E. & ICTV Report, C. 2022. ICTV virus taxonomy profile: rhabdoviridae 2022. *J Gen Virol,* June, 103(6). doi: 10.1099/jgv.0.001689. PMID: 35723908.

Whelan, S., Barr, J. & Wertz, G. 2004. Transcription and replication of nonsegmented negative-strand RNA viruses. *Curr Top Microbiol Immunol* 283, 61–119. doi: 10.1007/978-3-662-06099-5_3. PMID: 15298168.Springer Berlin Heidelberg.

Xu, K., Broder, C. C. & Nikolov, D. B. 2012. Ephrin-B2 and ephrin-B3 as functional henipavirus receptors. *Semin Cell Dev Biol,* 23, 116–23.

Yanagi, Y. 1995. Measles virus receptor CD46. *Uirusu,* 45, 159–64.

Yoshida, A., Kim, S. H., Manoharan, V. K., Varghese, B. P., Paldurai, A. & Samal, S. K. 2019. Novel avian paramyxovirus-based vaccine vectors expressing the Ebola virus glycoprotein elicit mucosal and humoral immune responses in guinea pigs. *Sci Rep,* 9, 5520.

14 Retroviruses

Satyendu Nandy[1] and Sachin Kumar[2]
Indian Institute of Technology Guwahati, Guwahati, Assam, India

INTRODUCTION

In 1908, the discovery of the chicken leukemia virus by scientists Bang and Ellenman started the journey for the discovery of retroviruses. In 1908, Peyton Rous stated, "Viruses can cause cancer." He found that the chickens suffering from leukemia were aged. So, he took the tumor, ground it, and passed it through a filter. After inoculating the filtrate, if it was still causing disease, then it was not a bacterial infection. That was the whole idea of the experiment. Then in 1911, the discovery of the Rous sarcoma virus occurred. He won the Nobel Prize 55 years later. Retroviruses are molecular biology's most studied group of viruses. They have some unique features. These groups of viruses are enveloped positive-sense ssRNA viruses containing RNA-dependent DNA polymerase (RdDp). In 1970, D. Baltimore and H. Temin showed that retroviruses encode RdDp and replicate through a DNA intermediate, which contradicts the central dogma theory. Later, they named it reverse transcriptase (RT). These viruses can integrate with the host chromosome and become host cellular genes. In 1975, Baltimore, Temin, and Dulbecco received the Nobel Prize for this discovery.

The first discovery of retroviruses started with the discovery of chicken sarcoma, a solid chicken tumor. Peyton Rous stated that tumors could be formed by the virus as well. Later, it became known as Rous sarcoma virus. Since then, many cancer-causing viruses have been isolated from animals and classified as oncornaviruses. In 1981, Robert Gallo et al. isolated human T-cell lymphotropic virus 1 (HTLV-1) from a patient sample of T-cell leukemia.

Long ago, there was a misconception that only homosexuals were susceptible to these viruses, especially HIV. HIV stands for human immunodeficiency virus. However, it is now known that anyone exposed to the virus can be affected. HIV is the root cause of acquired immunodeficiency syndrome (AIDS). HIV-1 can lead to AIDS, and it is found worldwide. HIV-2 is primarily found in West Africa and can cause AIDS-like symptoms in advanced conditions. From here onward, unless otherwise mentioned, all references to HIV will pertain to HIV-1.

CLASSIFICATION

In a broad sense, retroviruses are divided into two groups: (1) orthoretroviruses and (2) spumaviruses. Based on the mutations in the polymerase gene, orthoretroviruses are divided into α, β, δ, ϵ, γ, and lentiviruses. Spumaviruses are divided into bovine, feline, equine, prosimi (New World monkey), and simi (Old World monkey) spumaviruses. HIV can be classified in another way. There are three subfamilies of human disease-causing retroviruses: *Lentivirinae*, *Oncovirinae*, and Spumavirinae. There is another class of retrovirus named endogenous retroviruses. They may take up 8% of the human chromosome. Details of retroviruses are described in Table 14.1.

STRUCTURAL PROPERTIES

HIV preferentially infects CD4 T lymphocytes and destroys them, but they are necessary for our body's immune system. That makes the host highly susceptible to various opportunistic infections. Other cells, like macrophages and monocytes, can also be affected by this virus. In brief, the cells

DOI: 10.1201/9781003369349-14

TABLE 14.1
Classification of Retroviruses

Subfamily	Characteristics	Examples
HERVs	Retrovirus sequences that can bind to the human genome	Human placental virus
Lentivirinae	Slow disease progression; causes neurologic disorders and immunosuppression	HIV (HIV-1, HIV-2), visna virus (sheep), caprine encephalitis virus (goats)
Oncovirinae	Associated with cancer and neurologic disorders	—
B	Have eccentric nucleocapsid core in mature virus	Mouse mammary tumor virus
C	Have centrally placed nucleocapsid in mature virus	Human T-cell lymphotropic virus (HTLV-1, HTLV-2)
D	Have nucleocapsid core with cylindrical form	Mason–Pfizer monkey virus
Spumavirinae	Have foamy cytopathology	Human foamy virus

that have the CD4 proteins on their surface can be affected by HIV. They attach with some coreceptors to host cells. Common coreceptors used by HIV are CCR5 and CXCR4. According to their attachment to these cells, receptors, and coreceptors, its tropism is determined. They are called T-tropic (T cells), M-tropic (macrophages), R5 (CCR5), X4 (CXCR4), and R3 (CCR3) based on their tropism. T-cell-line-tropic (T-tropic) HIV-1 viruses use the chemokine receptor CXCR4 as a coreceptor, whereas macrophage-tropic (M-tropic) primary viruses use CCR5. However, some HIVs, when they attach to microglial cells, use CCR3 coreceptors (He et al., 1997).

HIV-1 and HIV-2 belong to the lentivirus subgroup, which refers to the slow infection caused by long incubation periods. They have a cylinder-shaped core with two positive-strand ssRNA inside it. Therefore, they are also called diploid viruses. The HIV-1 envelope has two specific glycoproteins, gp120 and gp41, whereas the HIV-2 envelope has gp130 and gp38. These glycoproteins are cleaved from gp160 (Figure 14.1).

GENOME AND REPLICATION OF HIV

The HIV genome is known to have one of the most complex structures among all other retroviruses. It has three typical retroviral genes: gag, pol, and env. Along with that, it has six additional genes. Among them, tat and rev are regulatory genes that help in replication. The other four genes—nef, vif, vpr, and vpu—are accessory genes unrelated to reproduction (Table 14.2).

The env gene encodes gp160, which is a precursor glycoprotein. It is cleaved to form gp120 and gp41. The gag gene encodes the internal core proteins of the virus. The essential core protein is p24. It is a medically necessary protein that acts as an antigen in the initial serological test to detect HIV antibodies inside the human body. The pol genes encode virion reverse transcriptase (RT), integrase, and protease. RT synthesizes DNA by using the RNA genome as a template. Integrase integrates viral DNA with host cellular DNA. Protease cleaves the various viral proteins. It is also medically essential to know the state of HIV because, without the function of proteases, the whole virus structure will not form in the body (Figure 14.2).

Reverse transcriptase is the source of the family name retroviruses. It is an RNA-dependent DNA polymerase. It helps in the transcription of the RNA genome into proviral DNA. It also works as a ribonuclease H enzyme. RNase H degrades viral RNA when it forms an RNA–DNA hybrid. This step is necessary to create the double-stranded proviral DNA. The integrase enzyme integrates

Retroviruses

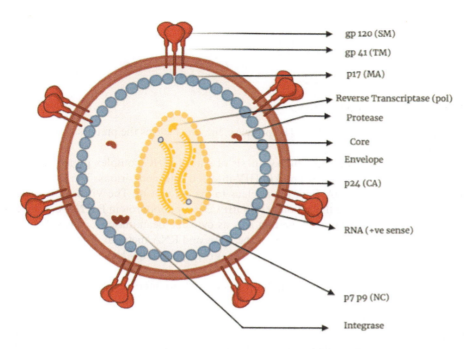

FIGURE 14.1 Cross-section view of the human immunodeficiency virus containing two positive-sense RNA strands inside the core. CA, capsid; MA, matrix; NC, nucleocapsid; SU, surface component; TM, transmembrane component of the envelope glycoprotein. (Figure created using BioRender.)

TABLE 14.2
Functions of Genes and Proteins in HIV

Genes	Proteins Encoded by Gene	Function of Proteins
1. Structural Genes (common to all retroviruses)		
Env	gp120	Attachment to CD4 proteins
	gp41	Fusion with host cells
Gag	p24, p7	Nucleocapsid
	p17	Matrix
Pol	Reverse transcriptase (RT)	Transcribes RNA genome to DNA
	Polymerase	Cleaves precursors polypeptides
	Integrase	Integrates viral DNA into host cells
2. Regulatory Genes of HIV		
Tat	tat	Transcription of the viral genome
Rev	rev	Transport late mRNA from the nucleus to cytoplasm
3. Accessory Genes of HIV		
Nef	nef	Decreases CD4 proteins and class I MHC proteins on the surface of infected cells
Vif	vif	Enhances infectivity by inhibiting the action of APOBEC3G
Vpr	vpu	In non-dividing cells, it transports viral core from the cytoplasm to nucleus
Vpu	vpr	Increase virus release from cells

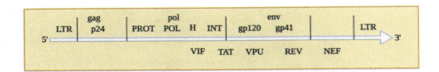

FIGURE 14.2 Genome structure of HIV.

proviral DNA into the host cell DNA. The protease enzyme cleaves the precursor polyproteins into functional viral polyproteins.

HIV proteins Tat and Nef repress the synthesis of class I MHC complex proteins. This reduces the capability of T cells to recognize and kill HIV-infected cells and virions. The Rev gene controls the transportation of late mRNA from the nucleus to the cytoplasm. The Vif protein increases the virus's infectivity by inhibiting the action of APOBEC3G (apolipoprotein B RNA-editing enzyme). APOBEC3G is an RNA-editing enzyme that acts as cytosine deaminase. It causes hypermutation in retroviral DNA, leading to the alteration of the DNA and RNA material of the virus, which is an important innate host defense against viruses. The Vif protein helps HIV evade the host immune system.

Envelope proteins gp120 and gp41 help the virus particle to interact with host cells. First, gp120 binds with the receptor, and then, gp41 binds with the coreceptor. The gene responsible for gp120 undergoes many mutations, resulting in antigenic variants. The V3 loop of gp120 is the most immunogenic region; most mutations happen here. For that reason, a vaccine against HIV is challenging to formulate. Antibody against nucleocapsid core protein is an important serological marker of HIV infection.

In the host cell, the cytoplasm virus replicates. From its positive sense, ssRNA forms an RNA–DNA complex with the help of host tRNA. Reverse transcriptase is used in that process. Then RNase H destroys that complex RNA, and from that cDNA, with the help of the polymerase enzyme, the virus makes its dsDNA. It goes to the nucleus and binds to any place in the host genome using the integrase enzyme. This virus genome with the host genome is called a provirus. The provirus makes its mRNA, and from mRNA, it makes its proteins. Proteins are made in polyprotein form, cleaved by proteases to form a functional protein. The important point is that gag and pol polyproteins are cleaved by viral protease, but the env polyprotein is cleaved by a cellular protease. Gag polyproteins form core proteins (p24) and matrix proteins (p17). Pol polyproteins include reverse transcriptase, integrase, and protease. Cell proteins like actin and tubulin help in virus movement inside the body, and cyclin T1 helps in some parts of the viral mRNA transcription process (Figure 14.3).

EPIDEMIOLOGY AND TRANSMISSION

HIV-1 is prevalent globally, but HIV-2 is significant in West African regions (Figure 14.4). HIV-2 is less transmissible than HIV-1 and closely related to simian immunodeficiency virus. There is a hypothesis that chimpanzees of West Africa were the source of HIV-1. Earlier, AIDS was known as GRID (gay-related immunodeficiency) because it was more frequent in homosexual men.

HIV can be transmitted by blood, sexual contact, and vertically. Drug abusers are prone to getting HIV infection. Some reports show that in some parts of the north-northeast of India, HCV (hepatitis C virus) and HIV are correlated. One patient infected with HIV eventually infects another one, or vice versa.

PATHOGENESIS AND IMMUNITY

The initial infection of the genital tract by HIV occurs in dendritic cells. The mucosa is lined by Langerhans cells. HIV is first found in the blood after 4–14 days of infection. It primarily affects

Retroviruses

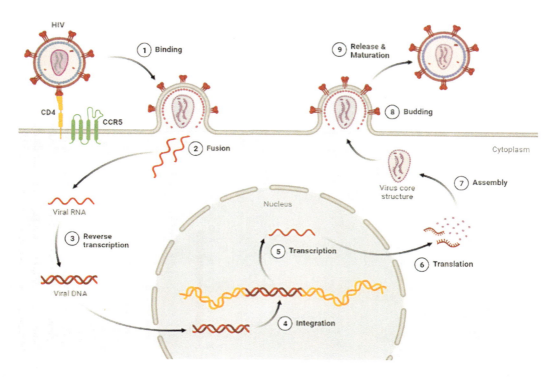

FIGURE 14.3 Replication cycle of HIV showing different steps in the formation process of a new virion.

CD4+ T helper cells (T_h). HIV can also act on a subset of CD4+ cells, such as T_h17 cells in the gastrointestinal (GI) tract. T_h17 cells mediate mucosal immunity in the GI tract. Destruction of the CD4+ cells results in the continuous decline of the host's cell-mediated immunity and makes them susceptible to many opportunistic infections, tumors, and lymphoma. Kaposi's sarcoma (HHV-8) is a common tumor in HIV-infected patients. T_h17 cells produce IL-17, which attracts neutrophils to bacterial infections. The absence of these cells causes blood infection by normal gut flora like *E. coli*. So, T_h cells are susceptible to HIV, but all other cells are permissible to HIV (Figure 14.5).

Some elite patients also have mutations in their CCR5 coreceptor. They are resistant to HIV. Approximately 1% of people of Western European ancestry have homozygous gene mutations. Resistance can also be obtained from the abundance of CXCR4 coreceptors. They will compete, and HIV cannot bind with the cells. There is no receptor-mediated endocytosis for HIV. Once they enter the host cells, they will dissolve because the membranes of mammals are rich in cholesterol. Cholesterol is called a beautification agent in cells. It gives fluidity to cells. The outer membrane of human cells does not express negatively charged lipids. The only negatively charged lipid molecule inside the inner cell membrane is phosphatidylserine. So far, no eukaryotic cells that express negatively charged lipid molecules have been reported. Bacterial cells are devoid of cholesterol and express negatively charged lipids in their outer membrane, which plays an antibacterial and antiviral role. Bacterial cells are devoid of lipid rafts. So, they are not infected by HIV.

The main immune response to HIV arises from CD8 lymphocytes. The envelope protein gp120 acts as a superantigen, causing the non-specific proliferation of T cells. Cytotoxic T cells lose their effectiveness because of the death of many CD4 helper T cells. This creates a scarcity of lymphokines (IL-2). HIV evades the immune response of host cells by binding with the host cell genome, undergoing mutations in the env gene, and increasing the production of tat and nef proteins, which downregulate the expression of class I MHC molecules.

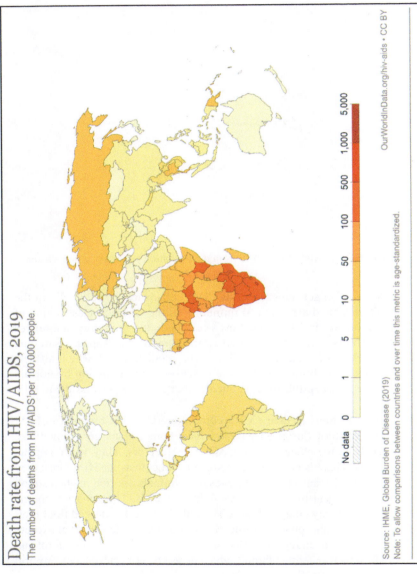

FIGURE 14.4 Death rates are high across sub-Saharan Africa. The large health burden of HIV/AIDS across sub-Saharan Africa is also reflected in death rates. Death rates measure the number of deaths from HIV/AIDS per 100,000 individuals in a country or region. In the interactive map, we see the distribution of death rates worldwide. Most countries have a rate of fewer than 10 deaths per 100,000; often much lower, below 5 per 100,000. Across Europe, the death rate is less than 1 per 100,000. Across sub-Saharan Africa, the rates are much higher. Most countries in the south of the region have rates greater than 100 per 100,000. In South Africa and Mozambique, it was once over 200 per 100,000. (Art from open source: ourworldindata.org/hiv-aids; Source: Global Burden Disease, 2019.)

Retroviruses

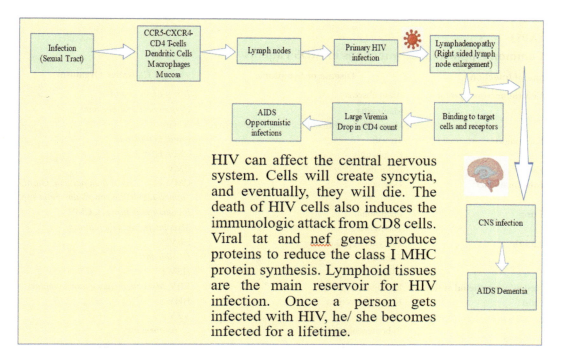

FIGURE 14.5 Pathogenesis of HIV.

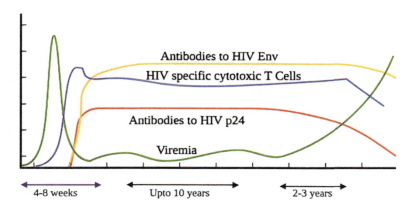

FIGURE 14.6 Time course of human immunodeficiency virus and the antibody responses.

CLINICAL FINDINGS

The clinical picture of HIV can be divided into three stages: (1) acute stage, (2) latent stage, and (3) immunodeficiency stage (Figure 14.6).

The acute stage occurs 2–4 weeks after infection. Symptoms include fever, lethargy, sore throat, rashes, leukopenia, and lymphadenopathy. The CD4 cell counts are usually average in this period, but a higher level of viremia will occur. However, the condition will resolve after two weeks of infection, which will generate a higher number of CD8 cells.

HIV antibodies will form inside the body 10–14 days after the infection. So, before that period, any serological test for detecting the antibody will give false negative results. However, the infected person can transmit HIV during this period. The PCR-based assay can be done for viral RNA to

TABLE 14.3

Common Opportunistic Infections in AIDS Patients

Site of Infection	Disease or Symptoms	Causative Organism
Central nervous system	Brain abscess	*Toxoplasma gondii*
	Meningitis	*Cryptococcus neoformans*
	Progressive multifocal leukoencephalopathy	JC virus
Esophagus	Thrush	*Candida albicans*
	Esophagitis	CMV, HSV-1
Eye	Retinitis	CMV
Intestinal tract	Diarrhea	CMV, *Cryptosporidium parvum, Giardia lamblia, Shigella* spp., *Salmonella* spp.
Lung	Pneumonia	*Pneumocystis jiroveci*, CMV
	Tuberculosis	*Mycobacterium tuberculosis*
Mouth	Hairy leukoplakia	EBV
	Thrash	*C. albicans*
	Ulcers	HSV-1
Reticuloendothelial system	Lymphadenopathy	EBV, *Mycobacterium avium complex*
Skin	Kaposi's sarcoma	HHV-8
	Zoster	VZV
	Subcutaneous nodules	*C. neoformans*

detect the virus. It is essential to determine the viral load in patients' bodies. After the initial viremia, a viral set point will occur in the body. This varies from person to person. The higher the set point, the higher the possibility of the disease progression to AIDS.

There is an established theory that an infected person can produce up to 10 billion copies of new virions each day, and every nucleotide of the HIV genome mutates daily. Eventually, patients develop signs and symptoms of opportunistic infections and sometimes neoplasms. A patient who has more than 10,000 copies of viral RNA per mL of plasma is more likely to progress to AIDS compared to a person who has less than 10,000 copies of viral RNA per mL of plasma.

CD4 cell count is another significant parameter. The average count for CD4 is 500 cells/μL. A person with this level of cell count or more usually remains asymptomatic. When the CD4 cell count starts decreasing, the rate of opportunistic infections starts increasing. If the CD4 cell count goes below 200 cells/μL, it is characterized as AIDS.

An untreated patient can survive up to 7–11 years with the latent period. During this period, a considerable amount of virions will be produced inside lymph nodes, and they become sequestered inside them. AIDS-related complex (ARC) is a typical latent stage feature that eventually progresses to AIDS.

The immunodeficiency stage is characterized by a decrease in the CD4 cell count and different kinds of opportunistic infections. Kaposi's sarcoma and pneumocystis pneumonia are common diseases in most AIDS patients. Other opportunistic diseases of that period are listed in Table 14.3.

LABORATORY DIAGNOSIS

HIV infection can be detected in three ways: (1) serologic determination, (2) virus isolation, and (3) measurement of viral nucleic acid or antigens.

HIV detection mostly depends on immunoassays. The detection of p24 antigen and antibody for HIV can be carried out by enzyme-linked immunosorbent assay (ELISA). A polymerase chain reaction (PCR), such as a nucleic acid amplification test (NAAT), can be performed for advanced detection.

Retroviruses 99

For early infection, a p24 antigen detection is a good approach. Some combo tests are also available to detect both the antibody and antigen. After that, PCR can be done to determine whether the infection is HIV-1 or HIV-2. Oral swabs can be used to perform an ELISA, but there is a possibility of getting false positive results from this test. A Western blot test can be done for the positive samples.

It is essential to calculate the viral load inside the body because it will help the clinician determine the treatment plan and predict how fast the disease will progress to AIDS. Moreover, CD4 cell counts and tests for drug resistance can also be done for HIV-infected patients.

TREATMENT

After the detection of HIV, it is suggested to start treatment as soon as possible. The therapy aims to (i) increase the CD4 cell count and restore the immune function of the body and (ii) decrease the viral load. It is important to note that treatment does not cure HIV completely, but it will control the virus replication inside the body; hence it will provide long-term suppression. However, the virus will show active replication if the drugs are stopped. Drug resistance is often seen in HIV strains. Hence, instead of single-drug therapy, a combination of drugs is used to treat HIV infection inside the body. It is known as anti-retroviral therapy (ART) or highly active anti-retroviral therapy (HAART).

Generally, drugs are chosen from those listed in Table 14.4. However, the choice of medicines depends on several factors such as the resistance of the HIV strains, CD4 cell count, viral load,

TABLE 14.4
Drugs against HIV

Entry inhibitors	Maraviroc
Fusion inhibitors	Enfuvirtide
Integrase inhibitors	Dolutegravir
	Raltegravir
Protease inhibitors	Atazanavir
	Darunavir
	Fosamprenavir
	Indinavir
	Nelfinavir
	Ritonavir
	Saquinavir
	Tipranavir
	Lopinavir/Ritonavir
Nonnucleoside RT inhibitors	Delavirdine
	Efavirenz
	Etravirine
	Nevirapine
	Rilpivirine
Nucleoside RT inhibitors	Abacavir
	Didanosine
	Emtricitabine
	Lamivudine
	Stavudine
	Zidovudine
Nucleotide RT inhibitors	Tenofovir

other opportunistic infections, patient's condition, and toxicity caused by the drug. Usually, three drugs are used to formulate a single drug, and the patient is given a single dose daily. This ART is also known as triple-drug therapy. Most commonly, combinations of RT inhibitors, integrase inhibitors, and protease inhibitors are used. ART is an expensive treatment, but millions of lives are saved with it. Despite its side effects, patients receiving ART live longer than patients without ART (Patel et al., 2008) (Figure 14.7).

Before initiating any treatment regimen, it is important to determine the viral resistance. It can be done by the genotyping method or phenotyping method. The genotyping method is the standard, but the phenotyping method should be used for any complex resistance pattern.

In the genotyping method, sequencing of RT and integrase is done to examine whether they are showing any resistance to the inhibitors of these gene products. A mutation is identified if the virus grows in the presence of its inhibitor drugs and coincides with clinical treatment failures.

In the phenotyping method, the recombinant virus grows in the presence of antiviral drugs. The relevant genes are cloned inside laboratory HIV strains. The drug concentration necessary to inhibit 50% of the viral replication (IC_{50}) is determined. The reference IC_{50} and patient virus IC_{50} ratio will indicate the fold resistance to the tested drug.

ENTRY AND FUSION INHIBITORS

From Table 14.4, maraviroc prevents the entry of HIV into the cells. It blocks the binding of the gp120 envelope protein of HIV to the coreceptor CCR5. However, this drug should be used only for R5 tropic viruses. Laboratory tests should be done to determine if the HIV strain is binding to CCR5. A Trofile assay is conducted for this test, wherein the patient's blood sample is tested to identify the tropism.

Enfuvirtide inhibits the fusion of the virus envelope with the cell membrane. It is a synthetic peptide-based drug that binds to gp41 and blocks HIV entry into cells.

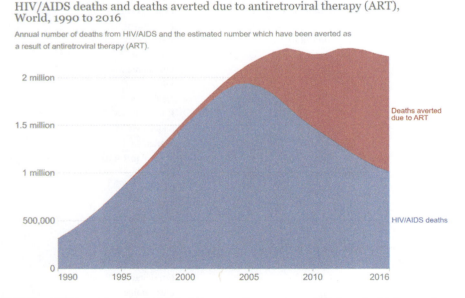

FIGURE 14.7 ART is a mixture of antiviral drugs used to treat people infected with human immunodeficiency virus (HIV). ART is essential in progressing against HIV/AIDS because it saves lives, allows people with HIV to live longer, and prevents new HIV infections. (Chart from open source: ourworldindata.org/hiv-aids; Source: UNAIDS.)

Integrase Inhibitors

Raltegravir is an integrase inhibitor that inhibits the integrase activity of HIV. This drug should be considered a last option and is recommended for patients who have already been treated with other ART but continue producing many virions inside the body.

Protease Inhibitors

Protease inhibitors are often given in combination with nucleoside RT inhibitors and are commonly used in ART. Sometimes a combination of protease inhibitors is also used, like lopinavir/ritonavir. Resistance to one protease inhibitor conveys resistance to all protease inhibitors.

Side effects of these drug treatments often result in abnormal fat deposition behind the neck region. It is one kind of lipodystrophy.

Nonnucleoside RT Inhibitors (NNRTIs)

Almost all the drugs from Table 14.4 are used to treat HIV. Nevirapine is often used to prevent vertical transmission of HIV from mother to fetus. Some of the NNRTIs can cause skin rashes and Stevens–Johnson syndrome.

Nucleoside/Nucleotide RT Inhibitors (NRTIs)

There are several NRTIs available. These drugs do not have a 3' hydroxyl group on the ribose ring and are therefore chain-terminating drugs. They do not cure the already infected or destroyed cells, but they can interfere with proviral DNA synthesis by RT, inhibiting HIV replication. These drugs have adverse effects on the body. Some can cause severe hypersensitivity reactions based on the gene structure of the patients. For instance, abacavir can cause hypersensitivity reactions in patients with the HLA-B1701 allele, characterized by fever, rash, malaise, and gastrointestinal and respiratory symptoms. Zidovudine can be used to prevent vertical transmission.

Vaccines against HIV

Vaccines are preventive and given to uninfected persons. Until now, all vaccine trials for HIV have shown poor results. Vaccine development is complex against HIV due to its mutability. HIV shows marked variation in its envelope antigens, which results in the neutralization of the vaccines.

Gene therapy approaches are being pursued against HIV-infected cells. It involves genetically altering target cells to become resistant to HIV. The main problem during vaccine development is the lack of animal models. Chimpanzees are susceptible to HIV. However, they are scarce. Moreover, they do not develop immunodeficiency.

Immune Reconstruction Inflammatory Syndrome (IRES)

HIV-infected patients, who are treated with a HAART regimen, often develop immune reconstruction inflammatory syndrome (IRES), and at the same time have a coinfection of HBV, HCV, *M. tuberculosis, M. avium complex, Cryptococcus neoformans, or Toxoplasma gondii*. In HIV infection, the patient's immune power is reduced due to the low count of CD4+ cells. However, when they start a HAART regimen, their immune system attempts to restore to normal condition by increasing the CD4+ cell count. This increases the patient's ability to produce an inflammatory response, exacerbating clinical symptoms. Therefore, before starting a HAART regimen, it is essential to treat any existing coinfection to avoid developing IRES (Table 14.5).

TABLE 14.5

Drugs for Advanced-Stage AIDS and Coinfection with Opportunistic Microbes

Drug	Infection Prevented
Azithromycin	*Mycobacterium avium* complex (MAC) infection
Clotrimazole	Oral thrush caused by *Candida albicans*
Fluconazole	Cryptococcal meningitis
Ganciclovir	Infection caused by HHV5 (cytomegalovirus)
Trimethoprim–sulfamethoxazole	*Pneumocystis* pneumonia, toxoplasmosis

PREVENTION

No vaccines for HIV have been available until now. So, preventive measures are necessary to avoid exposure to the virus, for instance, safe sex by using condoms, not sharing needles, screening of blood at blood banks, and discarding infected blood.

There are two kinds of prophylaxis available: preexposure prophylaxis (PrEP) and postexposure prophylaxis (PEP).

The drug emtricitabine (Truvada) can be given as a PrEP to individuals with a high risk of infection, such as homosexual males. People with high-risk nonoccupational exposure and needle-stick injury should be given PEP. The preferred PEP regimen consists of a combination of three drugs: emtricitabine, raltegravir, and tenofovir. Alternative combinations are also available. PEP should be administered as soon as possible after exposure to infection and continued for up to 28 days.

To avoid infection, children should be given ART if their mothers are HIV-infected. HIV-infected mothers should not breastfeed their babies. Clinical studies show there is less chance of getting an HIV infection if the delivery is done by cesarean section rather than vaginal delivery. For the treatment of HIV, current guidelines should always be followed.

ONCOGENIC RETROVIRUSES

Oncogenic retroviruses are initially RNA viruses. They create different kinds of leukemias, sarcomas, and lymphomas in many animals. These viruses are characterized by their properties to immortalize the infected cells. Transformation of cells results from overexpression or altered expression of genes. They can cause rapid proliferation where they carry oncogenes, create intermediate stages of gene activation by binding before an oncogene in the host (cis-activation), or activate the host oncogene from a distant place (transactivation).

The human oncoviruses include HTLV-1, HTLV-2, and HTLV-5, but only HTLV-1 is associated with the disease that causes adult T-cell leukemia (Kannian and Green, 2010).

HTLV-1 can spread in humans by blood transmission, sexual contact, and breastfeeding infected mothers. HTLV-1 can facilitate cell growth by integrating with the nearby cellular growth-controlling genes. Some African Americans, Southern Japanese, Australians, and Central Africans acquire HTLV-1 at birth. HTLV can cause ATLL, an aggressive (fast-growing) type of T-cell non-Hodgkin lymphoma. Malignant cells appear as flower cells. ELISA and RT-PCR tests can be done to detect HTLV.

ENDOGENOUS RETROVIRUSES

Different retroviruses have integrated into the chromosomes of humans and animals. They have a similarity of up to 8% of the human genome. These human endogenous retroviruses (HERVs) are replication-incompetent because of the deletion or insertion of termination codons or due to their poorly transcribed sequences. Moreover, our bodies have APOBEC proteins, which are apolipoprotein B-editing catalytic proteins that suppress the replication of endogenous retroviruses. During pregnancy, placental tissue forms syncytia to facilitate placental function. This is done by the activation of a HERV, which produces syncytin. Some HERVs are associated with prostate and other cancers, including multiple sclerosis. HERV type K (HERV-K) expression is usually inhibited in normal cells in healthy adults. However, HERV-K mRNA expression has been frequently documented to increase in tumor cells (Curty et al., 2020).

ACKNOWLEDGMENTS

Figures were created using Biorender.com, MS PowerPoint, and open-source images.

FURTHER READING

1) Review of Medical Microbiology and Immunology-A Guide to Clinical Infectious Disease-(15th Edition)- W. Levinson, P.C. Hong, E.A. Joyace, J. Nussbaum, B. Schwartz- Chapter 45: Human Immunodeficiency Virus
2) Jawetz, Melnick, Adelberg's Medical Microbiology (28th Edition)-Chapter 44: AIDS and Lentivirus
3) Medical Microbiology (9th Edition-P.R. Murray, K.S. Rosenthal, M.A. Pfaller)- Chapter 53: Retrovirus
4) Principals of Virology (4th Edition-Vol II; J. Flint, V.R. Racaniello, G.F. Rall, A.M. Skalka)- Chapter 7: Human Immunodeficiency Virus Pathogenesis.

REFERENCES

Curty, G., Marston, J. L., De Mulder Rougvie, M., Leal, F. E., Nixon, D. F. & Soares, M. A. 2020. Human endogenous retrovirus K in cancer: a potential biomarker and immunotherapeutic target. *Journal of Virology*, 12, 726.

He, J., Chen, Y., Farzan, M., Choe, H., Ohagen, A., Gartner, S., Busciglio, J., Yang, X., Hofmann, W. & Newman, W. 1997. CCR3 and CCR5 are co-receptors for HIV-1 infection of microglia. *Nature*, 385, 645–649.

Kannian, P. & Green, P. L. 2010. Human T lymphotropic virus type 1 (HTLV-1): molecular biology and oncogenesis. *Journal of Virology*, 2, 2037–2077.

Patel, K., Herná N, M. A., Williams, P. L., Seeger, J. D., Mcintosh, K., Dyke, R. B. V., Seage III, G. R. & PACTG Study Team. 2008. Long-term effectiveness of highly active antiretroviral therapy on the survival of children and adolescents with HIV infection: a 10-year follow-up study. *Clinical Infectious Diseases*, 46, 507–515.

15 DNA Viruses Containing Reverse Transcriptase

Sonali Sengupta and Baibaswata Nayak
Department of Gastroenterology, AIIMS, New Delhi, India

INTRODUCTION

Reverse-transcribing viruses use the reverse transcription process as a mode of replication. These viruses encode the reverse transcriptase (RT) enzyme for this process, and the enzyme is also known as RNA-dependent DNA polymerase, which uses RNA as a template to synthesize DNA. Normally, the transcription process uses DNA as a template to form messenger RNA (mRNA) that can express proteins, known as translation. Reverse transcription is a deviation from the central dogma of molecular biology, which states that there is a unidirectional flow of information from DNA to RNA and RNA to protein (Figure 15.1). The reverse-transcribing viruses fall into the Group VI and Group VII viruses as per the Baltimore classification. Group VI viruses have a single-stranded RNA genome that uses a DNA intermediate to replicate, whereas Group VII viruses have a double-stranded DNA genome that uses an RNA intermediate to replicate its genome. All the reverse-transcribing viruses belong to the class *Revtraviricetes*, which contains the virus families *Retroviridae*, *Metaviridae*, *Belpaoviridae*, *Psuedoviridae*, *Caulimoviridae*, and *Hepadnaviridae*. The reverse-transcribing DNA viruses belong to the families *Hepadnaviridae* and *Caulimoviridae*, whereas the other four families have RNA genomes as genetic material.

VIRUS CLASSIFICATION

Viruses can be classified based on their morphology, genomic nucleic acid type, mode of replication, infecting host type, and disease. The Baltimore classification system describes the ways viruses synthesize mRNA for protein production (Figure 15.2) and classifies all viruses into one of seven groups [1].

The viruses of these seven groups are as follows:

Group I (double-stranded DNA viruses, dsDNA): These viruses encapsidate dsDNA and use the classical route of information transmission, i.e., from DNA to RNA. The virus families *Polyomaviridae*, *Papillomaviridae*, *Adenoviridae*, *Herpesviridae*, *Iridoviridae*, and *Poxviridae* belong to this class.

Group II (single-stranded DNA viruses, ssDNA): These viruses encapsulate ssDNA, which is then replicated and expressed via a dsDNA intermediate. The virus families include *Parvoviridae* and *Circoviridae*.

Group III (double-stranded RNA viruses, dsRNA): These viruses package a dsRNA genome that is transcribed (where transcription is defined as mRNA synthesis) to produce the mRNA. The two virus families that belong to this class are *Reoviridae* and *Birnaviridae*.

Group IV (positive-sense single-stranded RNA viruses, ssRNA): These viruses have an ssRNA genome with positive (+) sense polarity, which can directly act as mRNA for the

104

DOI: 10.1201/9781003369349-15

DNA Viruses Containing Reverse Transcriptase

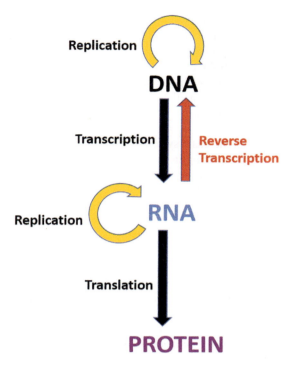

FIGURE 15.1 Central dogma of molecular biology.

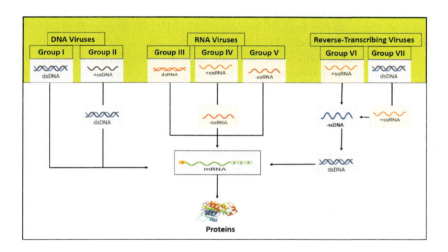

FIGURE 15.2 The Baltimore classification of viruses.

synthesis of viral proteins. The virus families of this class are *Caliciviridae*, *Picornaviridae*, *Flaviviridae*, *Togaviridae*, and *Coronaviridae*.

Group V (negative-sense single-stranded RNA viruses, ssRNA): These viruses also carry an ssRNA genome, but its polarity or sense is negative (–) or complementary to the mRNA. The genome needs to be transcribed later for viral protein production. The virus families *Filoviridae*, *Rhabdoviridae*, *Bunyaviridae*, *Orthomyxoviridae*, *Paramyxoviridae*, and *Arenaviridae* belong to this group.

Group VI (positive-sense ssRNA reverse-transcriptase viruses): These viruses carry a positive-sense RNA genome, which replicates via a DNA intermediate. The family *Retroviridae* belongs to this group.

Group VII (double-stranded DNA reverse-transcriptase viruses, dsDNA): These viruses package a dsDNA genome (although one of the strands is typically incomplete) that uses RT to replicate via an RNA intermediate. *Hepadnaviruses*, which include the hepatitis B virus (HBV), belong to this group.

The Baltimore classification was the basic system to classify viruses initially, but due to advances in genomic and metagenomic sequencing in the 21st century by modern next-generation sequencing techniques, the evolutionary landscape of the entire virosphere is now becoming discernible. The viruses of the Baltimore classification groups can be placed in the largest monophyletic taxa of viruses [1].

VIRUS TAXONOMY

The International Committee on Taxonomy of Viruses (ICTV) devises a scheme of virus classification considering the extensive information available about the viruses. The First ICTV Report, published in 1971, classified viruses into 43 virus groups, whereas the current Ninth Report describes a total of 87 virus families, including 2,284 virus species [2]. Until 2017, the ICTV developed a virus classification scheme with a five-rank hierarchy of viruses, including species, genus, subfamily, family, and order. This scheme mostly matches the five-rank structure of the Linnaean hierarchical structure used in the taxonomies of cellular organisms. Now the ICTV classifies formal viruses into a 15-rank hierarchy depending upon comparative sequence analyses of conserved genes and proteins [3]. This includes eight primary or principal ranks and seven secondary or their derivative ranks. The eight principal ranks are realm, kingdom, phylum, class, order, family, genus, and species [4]. As of now, there are 6 realms, 10 kingdoms, 17 phyla, 40 classes, 72 orders, 264 families, 2,818 genera, and 11,273 virus species [5].

The class *Revtraviricetes*, which includes all reverse-transcribing viruses, is divided into two orders: *Blubervirales* and *Ortervirales*. The reverse-transcribing DNA viruses of the order *Blubervirales* include the virus family *Hepadnaviridae* [6], whereas the reverse-transcribing DNA virus family *Caulimoviridae* belongs to the order *Ortervirales* [7]. The nackednaviruses, also reverse-transcribing DNA viruses, are identified from teleost fishes by metagenomic approaches [8], sharing many features with hepadnaviruses, and are provisionally classified into the *Nudnaviridae* family. The foamy viruses, or spuma viruses, carry both positive-sense ssRNA and dsDNA as genomes due to the process of late reverse transcription, which raises the question of whether these viruses should be considered DNA viruses or RNA viruses [9]. These viruses belong to the family *Retroviridae* and the subfamily *Spumaretrovirinae*. Normally, retroviruses have a positive-sense ssRNA genome with an intermediate stage of DNA. But these spuma or foamy viruses carry both RNA and DNA as genomes and share the characteristics that are more consistent with hepadnaviruses (DNA genome) rather than conventional retroviruses (RNA genome). The details of important viruses belonging to the reverse-transcribing DNA viruses are mentioned later.

FAMILY: *HEPADNAVIRIDAE*

The family *Hepadnaviridae* includes five genera with 18 species of viruses [6]. The five genera are *Orthohepadnavirus*, *Parahepadnavirus*, *Metahepadnavirus*, *Avihepadnavirus*, and *Herpetohepadnavirus*. Members of both *Metahepadnavirus* and *Parahepadnavirus* infect teleost fish, and most viruses in these genera have been identified by metagenomic approaches. Only white sucker hepatitis B virus (*Parahepadnavirus*) and bluegill hepatitis B virus (*Metahepadnavirus*) have been found as virions in infected hosts. Members of the genus *Herpetohepadnavirus* infect reptiles and frogs. The genus *Avihepadnavirus* comprises three species whose members infect birds, and the genus *Orthohepadnavirus* comprises 12 species whose members infect mammals. The family is placed within the realm *Riboviria*, order *Blubervirales*, because of homology between the hepadnavirus reverse transcriptase and the RNA-directed RNA polymerase of RNA viruses [10].

HEPATITIS B VIRUS (HBV)

There are five main hepatitis viruses that infect the liver and cause viral hepatitis, which is inflammation of the liver due to viral etiology. These viruses are hepatitis A virus (HAV, *Hepato, Picoranaviridae*), hepatitis B virus (HBV, *Orthohepadna, Hepadnaviridae*), hepatitis C virus (HCV, *Hepaci, Flaviviridae*), hepatitis D virus (HDV, *Deltavirus*, unclassified), and hepatitis E virus (HEV, *Orthohepe, Hepeviridae*). Other viruses associated with liver inflammation are cytomegalovirus, herpes simplex virus, coxsackievirus, and adenovirus.

CLINICAL SYMPTOMS

The hepatitis B virus is responsible for both acute and chronic infections of the liver. The clinical presentation of acute infection includes acute viral hepatitis (AVH), fulminant hepatitis, acute liver failure (ALF), acute-on-chronic liver failure (ACLF), and fulminant liver failure. Acute viral hepatitis is characterized by the sudden onset of symptoms like fever, nausea, and abdominal pain with jaundice or elevated serum aminotransferase levels. After convalescence, it may resolve automatically with viral clearance. Fulminant hepatitis is associated with hepatic encephalopathy. Acute liver failure is an acute loss of liver function or decompensation within 7–28 days of infection, characterized by events such as acute hemorrhage, severe abdominal infection, neurological impairment, and edema. ACLF is an acute deterioration of liver function in patients with underlying chronic liver disease following an acute insult or infection. Fulminant liver failure is the loss of liver function within one to six months of acute infection. Chronic HBV infection leads to chronic viral hepatitis, chronic liver disease (CLD), liver cancer or hepatocellular carcinoma (HCC), and end-stage liver disease or liver failure. HBV infection causes chronic viral hepatitis, which persists for more than six months. Chronic liver disease is the progressive destruction of liver parenchyma over a period of >6 months, leading to fibrosis, cirrhosis, and a chronic reduction in hepatic function. Chronic liver failure is characterized by poor synthetic, metabolic, and immunological functions and vascular compromise associated with ascites and portal hypertension.

TRANSMISSION

The main modes of HBV transmission are parenteral transmission, sexual transmission, horizontal transmission, and vertical transmission. Parenteral transmission pertains to the spread of an infectious agent that does not occur through the enteral or gastrointestinal tract route. Parenteral transmission of HBV occurs primarily through contaminated blood and blood products via the intravenous, intramuscular, and subcutaneous routes [11]. It can happen via transfusion or transmission through other nosocomial routes, such as dialysis, unsafe injections, needle-stick injuries, surgery, and dental procedures. Sexual transmission of HBV occurs primarily through sexual activity. Unvaccinated adults who engage in unprotected sex (vaginal, oral, or anal), have multiple sex partners, or have sex with partners who have a chronic hepatitis B infection are at an increased risk for transmission [12]. HBV infection is transmitted from mother to child through vertical transmission. It can occur in utero during pregnancy, perinatally during delivery, and postnatally during care and breastfeeding. Horizontal transmission of HBV that does not involve intravenous, sexual, or perinatal exposure mostly happens through mucosal contact with the persistent carrier's saliva, tears, tattooing, piercing, shaving in public, sharing needles at home or work, and other common situations that happen in places where the virus is endemic [13].

Epidemiology

Hepatitis B virus (HBV) infection is a major global health problem. According to the World Health Organization's (WHO) estimates for 2022, approximately 254 million people were living with chronic hepatitis B (CHB) infections, with 1.2 million new infections each year [14]. People with chronic CHB infection are most prevalent in the Western Pacific Region (97 million) and the African Region (65 million), respectively. There are 61 million CHB infections in the South-East Asia Region, 15 million in the Mediterranean Region, 11 million in the WHO European Region, and 5 million in the WHO Region of the Americas. Hepatitis B resulted in an estimated 1.1 million deaths in 2022 due to chronic liver disease, mostly from cirrhosis and hepatocellular carcinoma. The number of HBV-related deaths increased by 5.9% between 1990 and 2019 and by 2.9% between 2015 and 2019 [15]. In India, HBV infection is of intermediate endemicity (high >8%, intermediate 2%–8%, low <2%) and nearly 2–4% of the population (about 40 million people) are CHB carriers [16]. Most of this hepatitis B global burden can be attributed to mother-to-child transmission during pregnancy or shortly after birth and horizontal household transmission.

MOLECULAR BIOLOGY OF HBV

Virion Morphology and Genomic Organization

HBV is an enveloped, spherical virus with a diameter of approximately 42 nm. The infectious virion consists of a viral genome, a nucleocapsid core, and an HBsAg-made lipid bilayer envelope (Figure 15.3). The infectious virion is called the Dane particle, and other subviral particles of HBV are also abundant in infected patients' serum, but these are devoid of the nucleocapsid core and genome [17]. These occur both in spherical forms of particle size of 20–25 nm in diameter and in filaments of 20–22 nm in diameter and of variable length [18].

The HBV viral genome is a 3.2 kb long relaxed circular DNA (rcDNA) molecule that has a full minus strand and a partial positive-sense strand [19]. Base-pairing of an overlap between the 5′ ends of the plus and minus strands of DNA leads to the formation of a circularized DNA molecule. A covalent bond links the 5′ end of the minus strand to a tyrosine (Y) hydroxyl residue in the terminal

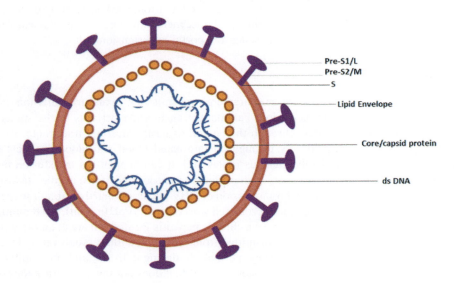

FIGURE 15.3 Structure of HBV. Diagram of hepatitis B virus (HBV) depicting surface proteins (large [Pre-S1/L], middle [Pre-S2/M], and small [S] surface proteins), lipid envelope, core or capsid protein, and partial ds (double-stranded) DNA genome. (Generated with BioRender, www.biorender.com.)

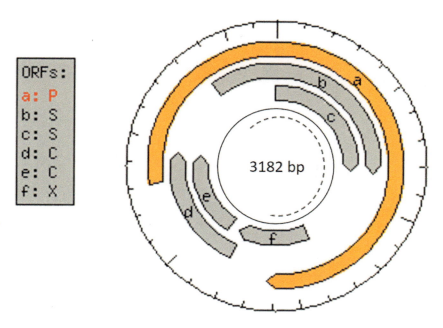

FIGURE 15.4 Schematic of the HBV genome showing overlapping open reading frames.

protein of HBV polymerase. A 19-nt-long 5′-capped RNA region of the pregenomic RNA (pgRNA) forms the 5′ end of the plus DNA strand. The genome contains either wholly or partially overlapping four main open reading frames (ORFs): polymerase (P), core (C), surface (S), and X proteins that encode seven proteins (P, C, preCore/HBe, X, L, M, S).

These overlapping ORFs within the HBV circular genome are depicted in Figure 15.4. The longest ORF encodes the viral polymerase (P). This is a polyprotein that is comprised of three subdomains: the terminal protein domain, priming the minus-strand DNA synthesis; the reverse transcriptase, including the DNA-dependent DNA polymerase for plus-strand synthesis; and the RNase H domain that is located at the carboxy-terminus. The other overlapping open reading frame encodes surface (S) proteins. Three co-carboxy-terminal surface (HBs) proteins are encoded by this ORF, and these proteins are large (L-HBs), middle (M-HBs), and small (S-HBs) proteins. The core ORF encodes both precore/HBeAg and the structural core protein. The HBeAg is a nonstructural form of the core proteins that are efficiently secreted. The X protein, encoded by X-ORF, whose function is still not fully understood, is believed to be involved in HBV-associated carcinogenesis.

HBV Life Cycle

The HBV life cycle (Figure 15.5) begins with virion attachment to the cell (step 1), followed by virus–cell envelope fusion and viral capsid release in the cell cytoplasm (step 2). HBV attaches to the receptor heparan sulfate proteoglycans on the surface of hepatocytes [20–22], which sets up conditions for virus entry by a high-affinity interaction of the preS-domain of LHBs with the HBV-specific receptor sodium taurocholate co-transporting polypeptide (NTCP) [23, 24]. Following attachment, clathrin-mediated endocytosis of the HBV virion occurs [25, 26] with the cytoplasmic release of the HBV nucleocapsid and its transportation to the cell nucleus through the nuclear pore complex [27]. The entry of rcDNA into the nucleus (step 3) is followed by the removal of the polymerase protein, which is covalently attached to the rcDNA. The repair of the rcDNA results in the formation of covalently closed circular (ccc) DNA that usually gets archived in the nuclei of HBV-infected cells [28]. The cccDNA is the extrachromosomal form of HBV DNA. The linear form of

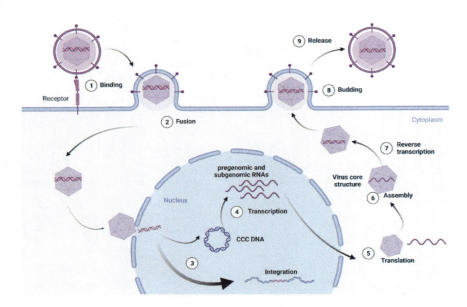

FIGURE 15.5 Schematic of the HBV life cycle in a liver cell depicting steps 1–9. (Generated with BioRender, www.biorender.com.)

HBV DNA can also be integrated into the host DNA genome at double-stranded DNA breaks [29, 30]. Transcription from cccDNA leads to pregenomic (pgRNA) and subgenomic RNA production (step 4), and these mRNAs get transferred to the cytoplasm (step 4). Translation of pregenomic and subgenomic RNAs leads to the formation of viral structural and non-structural proteins (step 5). Both core and polymerase proteins are synthesized from the pgRNA transcript. The pgRNA also serves as a template for negative-sense DNA synthesis by reverse transcription. The core protein assembles pgRNA and DNA polymerase to form nucleocapsids (step 6) for maturation by reverse transcription-mediated DNA synthesis and DNA replication. Mature nucleocapsids contain the newly synthesized partially double-stranded rcDNA, with the polymerase still bound to the 5′ end of the negative-sense DNA strand (step 7). Early on in infection, the mature nucleocapsid can shuttle back to the nucleus to replenish the cccDNA pool. Later, it can be used for progeny virion formation when enough HBsAg accumulates in the cytoplasm [31]. After assembly and genome maturation, the core protein exhibits an affinity for intracellular membranes, which incorporate large surface protein molecules [32]. For the secretion of enveloped virions, an excess of small surface proteins is necessary [32, 33]. During secretion, virions and accompanying virus-like particles move from the endoplasmic reticulum (ER) via the Golgi apparatus to the cell surface for budding (step 8) [32]. During the envelopment of core particles and secretion, the large surface protein molecule refolds in the pre-S domain from the internal face of the envelope to the surface. Human hepatoma cell lines such as HepG2 and Huh7 express a suitable ratio of all HBV proteins to allow for virion assembly and secretion [34]. HBV budding involves two endocytic host proteins—the ubiquitin-interacting adaptor 2-adaptin and the Nedd4 ubiquitin ligase—as well as the cellular machinery that generates internal vesicles of multivesicular bodies (MVBs) for transport and release (step 9) [35].

TRANSCRIPTION AND TRANSLATION OF VIRAL GENES

The HBV genome has genes that code for proteins and all the regulatory elements. These include two enhancers (Enh1 and Enh2); four promoters (core, S1, S2, and X); and polyadenylation, encapsidation (ε), and replication (DR1 and DR2) signals that are built into the ORFs. The rcDNA is repaired in the nucleus using host cellular repair mechanisms by completion of the plus strand, removal of 5′

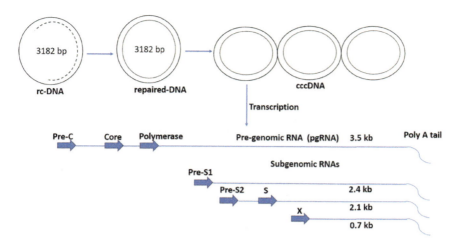

FIGURE 15.6 Transcription of HBV RNA.

terminal structures, including terminal protein in the minus strand, and oligoribonucleotide primer in the plus strand. The process then involves ligating the respective strands to produce supercoiled cccDNA [36]. The cccDNA (negative-strand nucleotide sequence) acts as a template for the transcription of genomic and subgenomic RNAs (Figure 15.6). These RNAs are 5' cap and 3' polyadenylated at a common polyadenylation signal located in the core ORF. Transcription of four HBV ORFs from double-stranded cccDNA into RNAs is a highly regulated process that uses enhancer II (EN2)/basal core promoter (BCP), large surface antigen, major surface antigen, X gene promoters, and enhancer I (EN1) [34]. The pgRNA is greater than the genome length (~3.5 kb) and has redundant ends that are essential for genome replication and act as a template for the synthesis of the viral DNA negative strand through reverse transcription. It translates the precore, core, and polymerase proteins. The other three subgenomic RNA transcripts are 2.4 kb, 2.1 kb, and 0.7 kb in length, which are translated, respectively, into the envelope preS1 (L) protein, the preS2 (M) and HBsAg proteins, and the X protein [37].

PRECORE/CORE PROTEIN

The precore/core ORF starts with two initiation (AUG) codons at the 1812 and 1899 nt positions. Translation from the first initiation codon produces a precore protein of 212 aa in length and a protein size of 25 kDa. The translation of the second initiation codon results in the production of the 21 kDa HBV core antigen (HBcAg), composed of 183 amino acids. The precore protein has an N-terminal signal sequence (−29 to 1 aa) and a C-terminal arginine-rich domain (149–183 aa). Complete post-translational modification produces HBeAg, whereas incomplete post-translational processing produces precore-related protein (P22cr) (Figure 15.7).

The translation initiation at the precore start or first codon suppresses the use of downstream-located AUGs [38]. The primary precore translation product (p25) is further processed in the ER and secreted as HBeAg through the ER–Golgi apparatus secretory pathway. For HBeAg production, two proteolytic cleavage events occur. The N-terminal region contains a 29-amino-acid-long signal sequence, but the cellular signal peptidase first cleaves a 19 aa long hydrophobic signal peptide during precore translocation into the ER lumen, reducing p25 to the intermediate size p22 [39]. The p22 is thus identical to HBc except for the 10 aa N-terminal extension, which causes the folding of the p22 protein to differ from that of the core protein, allowing dimerization but no further assembly of these dimers into capsids. The second proteolytic processing of p22 protein occurs in the trans-Golgi apparatus, where a 30 aa long arginine-rich domain is cleaved by proprotein convertases such as furin in the secretory pathway to generate multiple species of HBeAg due to heterogeneity in C-terminal processing [40]. The cleavage of the 30 aa C-terminal region produces

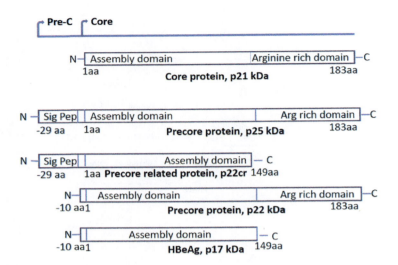

FIGURE 15.7 Precore/core protein translation.

another precore-derived species, p22cr, which is distinct from the intracellular p22 and has an intact N-terminal signal sequence [41]. The secretory HBeAg protein acts as an immunomodulatory molecule and mainly induces immune tolerance at the T-cell level or acts as a T-cell tolerogen. This immune regulation may contribute to chronicity during perinatal infections and prevent severe liver injury during adult infections [42]. The core protein (HBcAg) N-terminal region (assembly domain) is essential for self-assembly, whereas the C-terminal arginine-rich region is important for interaction with the pgRNA/reverse transcriptase complex and packaging. These core proteins self-assemble to form an icosahedral capsid. There are two types of capsids that have been observed: one with an icosahedral symmetry of T=4, about 34 nm in diameter, has 240 copies of capsid protein; and the other with an icosahedral symmetry of T=3, about 30 nm in diameter, has 180 copies of core particle [43]. The phosphorylation of capsid protein is critical for nuclear localization because it binds to the nuclear pore complex and gets imported to the nucleus. The capsids that contain the mature viral genome can release the viral DNA and capsid protein into the nucleus for more genomes for transcription, or they can bud through the endoplasmic reticulum to provide new virion.

POLYMERASE (P) PROTEIN

The viral polymerase (P or Pol) is an 832 aa long, large (94 KDa) multifunctional protein that is encoded by 80% of the viral genome. The protein consists of four domains: terminal protein (TP) or primase, spacer, polymerase/reverse-transcriptase (RT), and ribonuclease H (RH) domain (Figure 15.8). The TP domain at the N-terminus is needed for epsilon binding, pgRNA packaging,

FIGURE 15.8 Schematic of HBV polymerase depicting different domains: terminal protein (1–178 amino acids [aa]); spacer (179–336 aa); reverse transcriptase (337–680 aa); RNaseH (681–832 aa). Catalytic YMDD amino acids between 538 and 541 aa. The reverse transcriptase domain contains various subdomains named A, B, C, D, E, F, G. Tyrosine, Y; methionine, M; aspartic acid, D.

DNA Viruses Containing Reverse Transcriptase

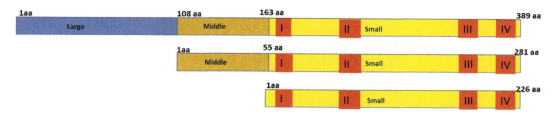

FIGURE 15.9 Schematic depicting co-terminal envelope proteins. Large: 389 aa, Middle: 281aa, Small: 226 aa Transmembrane domains are represented as I, II,III, IV on the surface polypeptide.

and protein priming to produce negative-sense DNA strand synthesis. The "spacer" domain can be mutagenized without disrupting its function. Three cysteine residues located in the carboxy terminal region of the spacer, as well as one additional cysteine residue in the N-terminus of the RT domain, are required for RNA packaging.

The polymerase domain has both RNA-dependent polymerase activity (RT) and DNA-dependent polymerase activity. The HBV RT domain has shown significant homology with retroviral RT, which includes boxes A through E along with boxes F (or box II) and G (or box I). The conserved catalytic active site for polymerase activity is the tyrosine–methionine–aspartate–aspartate (YMDD) motif localized in box C. The polymerase protein is crucial for genome replication. The RNaseH domain is important for RNaseH activity by coordinating metal ion binding. The negatively charged residues containing the DEDD motif are important for HBV replication, pgRNA packaging, and RNaseH activity. Both TP and RT domain interactions are required for epsilon binding and protein priming.

ENVELOPE/SURFACE PROTEINS

Three HBV envelope proteins (L, M, and S) are expressed from two subgenomic RNAs that are transcribed from the preS/S ORF. All these three proteins share a common carboxy (C) terminal domain and a unique N-terminal domain, such as the preS1 domain for the L protein, the preS2 domain for the M protein. and the S domain for the S protein (Figure 15.9). There is no signal sequence in the preS1 domain of the L protein, therefore not translocated to the ER to follow the secretory pathway and is restricted to the cytosolic side of the ER and not N-glycosylated. The signal sequences or transmembrane domains (TMDs) I and II of the S domain are responsible for translocation into the ER membrane, but the entire pre-S domain is too long to be translocated into the ER lumen during protein synthesis [44]. The central hydrophilic part of the small surface domain is in the lumen of the ER and is more efficiently N-glycosylated than the S domain of L and M surface proteins [34]. PreS1 epitopes displayed on the cytosolic side of the viral membrane act as ligands for interaction with core particles during the assembly of the viral envelope.

X PROTEIN

The HBx gene, the smallest ORF of the HBV genome, encodes a 154-amino acid regulatory protein with a molecular weight of 17 kDa. It is named X protein due to the lack of sequence homology to any existing protein [45]. HBx predominantly accumulates in the cytoplasm when highly expressed, whereas it localizes to the nucleus at low expression levels [46, 47]. It has been proposed that HBx is multifunctional owing to its negative regulatory/anti-apoptotic N-terminal and transactivator C-terminal, in addition to being the only regulatory protein encoded by the virus. During HBV viral infection, HBx acts on cellular and viral promoters and enhancers, driving transactivation through protein–protein interactions [48, 49]. HBx protein is implicated in hepatocellular carcinoma [45, 50, 51].

HBV Replication

HBV replication begins with reverse transcription of pgRNA to dsDNA production. Reverse transcription occurs largely in intact nucleocapsids, which is an extreme variation for reverse-transcribing viruses as there is a space restriction imposed on the replicating complex inside the geometrically defined capsid lumen. It appears that no full-length DNA can form outside of capsids or in the absence of core proteins. Thus, the capsid is considered a dynamic replication machine [52]. When sufficient amounts of core and polymerase proteins are translated, these proteins assemble together with the pgRNA and cellular proteins [53, 54]. Encapsidation occurs only when the polymerase interacts with the epsilon sequence (Figure 15.10i) at the 5′ end of the pgRNA [55]. Binding of the polymerase/RT alters the structure of epsilon [56], and this binding to epsilon is also required for the activation of polymerase [57, 58]. The pgRNA epsilon (ε) is a stem–loop structure, and the epsilon bulge or apical loop consists of three nucleotides (UUC) from where the initiation of reverse transcription happens, so the epsilon is considered the replication origin [59, 60]. The polymerase orients itself around the epsilon bulge.

The TP domain is separated from the RT and RNaseH domains by a highly variable spacer, which provides a specific tyrosine (Y) residue to which the first nucleotide of minus-strand DNA becomes covalently linked (Y63 in HBV TP), and synthesis of the oligonucleotide 5′-TGAA-3′ occurs. The oligonucleotide bound to TP must specifically be translocated or slid to the DR1*, nearly 3 kb apart from 5′ epsilon (Figure 15.10i). The primer translocation process involves remodeling of the

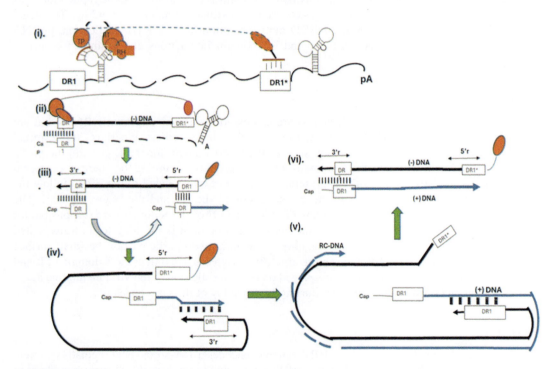

FIGURE 15.10 Steps in HBV DNA replication: (i) Reverse transcriptase orients itself on the RNA pregenome epsilon bulge. (ii) Reverse transcription of pre-genomic RNA to minus (−) strand DNA is associated with concomitant degradation of pre-genomic RNA. (iii) RNA primer translocation from the 3′ end of (−)DNA to the 5′ end of (−)DNA, with the beginning of plus (+) strand DNA synthesis. (iv) Circularization of minus-strand DNA, with continued synthesis of plus strand DNA via a template switch to the 3′ end of minus-strand DNA. (v) Continuation of synthesis of (+)DNA on (−)DNA template, leading to the formation of relaxed circular double-stranded DNA in 90% of the virions. (vi) Variant conformation of double-stranded linear DNA form may also be observed [2]. (Modified from Beck and Nassal, 2007.)

P-epsilon complex, the priming Y residue of TP giving way to the growing DNA oligonucleotide, then epsilon must be replaced by DR1* as the template [52]. Extension of 5'-TGAA-3' occurs for synthesis of minus-strand DNA. The end product of negative-sense strand DNA synthesis is a unit-length DNA copy of the pgRNA starting from its 5' end to the UUCA motif in the 3' DR1* with 10 nt, terminal redundancy. The pgRNA is degraded concomitantly as minus-strand DNA proceeds by RNase H activity of RT (Figure 15.10ii). RNA primer translocation from the 3' end of negative sense-DNA to the 5' end of negative-sense DNA occurs with the beginning of plus-strand DNA synthesis (Figure 15.10iii). The capped 5' RNA oligonucleotide functions as a primer for plus-strand DNA synthesis. Extension of the RNA from its original position gives rise to a double-stranded linear DNA which occurs to a small percentage, about 10% in all hepadnaviruses [61]. For relaxed circular (RC)-DNA formation, the RNA primer must be transferred to the 3' proximal DR2. Further elongation requires a third template switch, i.e., circularization. During this step, the growing plus-strand DNA end is transferred from 5'r to 3'r on the minus-strand DNA template, from where it can be further extended to yield relaxed circular (RC)-DNA (Figure 15.10iv–vi).

HBV GENOTYPES

Due to the lack of proofreading activity of its reverse transcriptase, the mutation rate of HBV is very high, with an estimated nucleotide replacement rate of 1.4 to 3.2×10^{-5} per site per year. The high error rate of HBV reverse transcriptase during viral replication results in frequent nucleotide substitutions, leading to genotype, subgenotype, and quasispecies diversity [62]. In 1988, Okamoto et al. suggested that HBV could be divided into four genotypes based on genome nucleotide variation that is greater than 8%. Since then, at least ten genotypes (A–J) have been identified [62].

HEPATITIS B VIRUS DIAGNOSIS

HBV infection is diagnosed using both serological and molecular diagnostic tests. Various serological assays can detect virus-specific antigens (HBsAg, HBeAg) and antibodies that appear during (anti-HBc IgM, anti-HBe IgG) and after (anti-HBs, total anti-HBc, and anti-HBe) HBV infection. Molecular diagnostic tests include quantitative or qualitative detection of HBV DNA using nucleic acid detection-based tests. The detection of HBsAg, which can be the virus envelope, with or without virus DNA, i.e., the virion or HBs spheres and filaments in the serum indicates current HBV infection, and second surface antigen test positivity after six months indicates chronic infection. Detection of anti-HBs antibody indicates either recovery from HBV infection or immunization against HB infection either by HB vaccine or prior infection. Detection of anti-HBc antigens relates to exposure to HBV, and infection may be acute, chronic, or resolved. Detection of HBeAg in serum indicates active viral replication, and appearance of anti-HBe indicates resolution of the infection.

HEPATITIS B TREATMENT

Chronic hepatitis B (CHB) patients require long-term antiviral therapy to prevent disease progression to liver cirrhosis, hepatic failure, hepatocellular carcinoma, and death. Both interferon (IFN)-based immunomodulatory therapy and nucleos(t)ide analogues (NUCs or NAs)-based antiviral therapies are indicated for the treatment of CHB patients [63]. The NAs for CHB treatments are classified into three main groups according to their chemical structure. These include L-nucleoside analogues (lamivudine, emtricitabine, telbivudine, torcitabine, and clevudine), acyclic nucleoside phosphonates (adefovir and tenofovir), and deoxyguanosine analogues (entecavir and abacavir). Out of these NAs, entecavir and tenofovir are the most potent drugs with high genetic resistance [64]. The development of drug resistance begins with mutations in the polymerase/RT gene, followed by viral and biochemical breakthroughs with clinical deterioration leading to the loss of favorable effects obtained by NA therapy [65]. Major mutational patterns conferring nucleoside/nucleotide analogue resistance include the combinations rtL180M/rtM204 (I/V) (for lamivudine, entecavir, telbivudine, and clevudine) and rtA181V/rtN236T (for adefovir and tenofovir). The mutations in the HBV RT region (rtA181, rtM204, and rtN236) have been reported to be associated with major primary drug resistance and are responsible for reduced treatment susceptibility resulting from antiviral agents [66].

HBV VACCINE

Safe and effective HBV vaccines have been available since 1982. The HBV vaccine consists of the HBV surface antigen as an immunogen. These vaccines protect against different HBV genotypes (A–H) and can provide prolonged immunity (~up to 25 years). The WHO recommends universal infant immunization with at least three doses of the hepatitis B vaccine, each separated by at least four weeks. The first dose of the hepatitis B vaccine should be given preferably within 24 hours after birth. Unvaccinated individuals should be vaccinated according to a 0-, 1-, and 6-month schedule for better protective immunity [67].

REFERENCES

1. Koonin, E.V., M. Krupovic, and V.I. Agol, The Baltimore classification of viruses 50 years later: how does it stand in the light of virus evolution? *Microbiol Mol Biol Rev*, 2021. **85**(3): p. e0005321.
2. Davison, A.J., Journal of general virology - introduction to ICTV virus taxonomy profiles. *J Gen Virol*, 2017. **98**(1): p. 1.
3. Siddell, S.G., et al., Virus taxonomy and the role of the International Committee on Taxonomy of Viruses (ICTV). *J Gen Virol*, 2023. **104**(5): p. 001840. doi: 10.1099/jgv.0.001840. PMID: 37141106; PMCID: PMC10227694
4. International Committee on Taxonomy of Viruses Executive, C., The new scope of virus taxonomy: partitioning the virosphere into 15 hierarchical ranks. *Nat Microbiol*, 2020. **5**(5): p. 668–74.
5. Walker, P.J., et al., Recent changes to virus taxonomy ratified by the international committee on taxonomy of viruses (2022). *Arch Virol*, 2022. **167**(11): p. 2429–40.
6. Magnius, L., et al., ICTV virus taxonomy profile: Hepadnaviridae. *J Gen Virol*, 2020. **101**(6): p. 571–2.
7. Teycheney, P.Y., et al., ICTV virus taxonomy profile: caulimoviridae. *J Gen Virol*, 2020. **101**(10): p. 1025–6.
8. Lauber, C., et al., Deciphering the origin and evolution of hepatitis B viruses by means of a family of non-enveloped fish viruses. *Cell Host Microbe*, 2017. **22**(3): p. 387–99 e6.
9. Yu, S.F., M.D. Sullivan, and M.L. Linial, Evidence that the human foamy virus genome is DNA. *J Virol*, 1999. **73**(2): p. 1565–72.
10. Koonin, E.V., et al., Global organization and proposed megataxonomy of the virus world. *Microbiol Mol Biol Rev*, 2020. **84**(2): p. e00061-19. doi: 10.1128/MMBR.00061-19. PMID: 32132243; PMCID: PMC7062200.
11. Berkley, S., Parenteral transmission of HIV in Africa. *AIDS*, 1991. **5**(Suppl 1): p. S87–92.
12. di Filippo Villa, D. and M.C. Navas, Vertical transmission of hepatitis B virus-an update. *Microorganisms*, 2023. **11**(5).: p. 1140. doi: 10.3390/microorganisms11051140. PMID: 37317114; PMCID: PMC10221798.
13. Davis, L.G., D.J. Weber, and S.M. Lemon, Horizontal transmission of hepatitis B virus. *Lancet*, 1989. **1**(8643): p. 889–93.
14. Easterbrook, P.J., et al., WHO 2024 hepatitis B guidelines: an opportunity to transform care. *Lancet Gastroenterol Hepatol*, June 2024. **9**(6): p. 493-495. doi: 10.1016/S2468-1253(24)00089-X. Epub 2024 Apr 10. PMID: 38614110.
15. Collaborators, G.B.D.H.B., Global, regional, and national burden of hepatitis B, 1990–2019: a systematic analysis for the Global Burden of Disease Study 2019. *Lancet Gastroenterol Hepatol*, 2022. **7**(9): p. 796–829.
16. Tandon, B.N., S.K. Acharya, and A. Tandon, Epidemiology of hepatitis B virus infection in India. *Gut*, 1996. **38**(Suppl 2): p. S56–9.
17. Dane, D.S., C.H. Cameron, and M. Briggs, Virus-like particles in serum of patients with Australia-antigen-associated hepatitis. *Lancet*, 1970. **1**(7649): p. 695–8.
18. Kaplan, P.M., et al., DNA polymerase associated with human hepatitis B antigen. *J Virol*, 1973. **12**(5): p. 995–1005.

19. Tsukuda, S. and K. Watashi, Hepatitis B virus biology and life cycle. *Antiviral Res*, 2020. **182**: p. 104925.
20. Schulze, A., P. Gripon, and S. Urban, Hepatitis B virus infection initiates with a large surface protein-dependent binding to heparan sulfate proteoglycans. *Hepatology*, 2007. **46**(6): p. 1759–68.
21. Leistner, C.M., S. Gruen-Bernhard, and D. Glebe, Role of glycosaminoglycans for binding and infection of hepatitis B virus. *Cell Microbiol*, 2008. **10**(1): p. 122–33.
22. Sureau, C. and J. Salisse, A conformational heparan sulfate binding site essential to infectivity overlaps with the conserved hepatitis B virus a-determinant. *Hepatology*, 2013. **57**(3): p. 985–94.
23. Ni, Y., et al., Hepatitis B and D viruses exploit sodium taurocholate co-transporting polypeptide for species-specific entry into hepatocytes. *Gastroenterology*, 2014. **146**(4): p. 1070–83.
24. Yan, H., et al., Sodium taurocholate cotransporting polypeptide is a functional receptor for human hepatitis B and D virus. *Elife*, 2012. **1**: p. e00049.
25. Herrscher, C., P. Roingeard, and E. Blanchard, Hepatitis B virus entry into cells. *Cells*, 2020. **9**(6).
26. Umetsu, T., et al., Inhibitory effect of silibinin on hepatitis B virus entry. *Biochem Biophys Rep*, 2018. **14**: p. 20–25.
27. Rabe, B., et al., Nuclear entry of hepatitis B virus capsids involves disintegration to protein dimers followed by nuclear reassociation to capsids. *PLoS Pathog*, 2009. **5**(8): p. e1000563.
28. Tu, T., H. Zhang, and S. Urban, Hepatitis B virus DNA integration: in vitro models for investigating viral pathogenesis and persistence. *Viruses*, 2021. **13**(2): p. 180. doi: 10.3390/v13020180. PMID: 33530322; PMCID: PMC7911709.
29. Bill, C.A. and J. Summers, Genomic DNA double-strand breaks are targets for hepadnaviral DNA integration. *Proc Natl Acad Sci U S A*, 2004. **101**(30): p. 11135–40.
30. Tu, T., et al., HBV DNA integration: molecular mechanisms and clinical implications. *Viruses*, 2017. **9**(4): p. 75. doi: 10.3390/v9040075. PMID: 28394272; PMCID: PMC5408681.
31. Lentz, T.B. and D.D. Loeb, Roles of the envelope proteins in the amplification of covalently closed circular DNA and completion of synthesis of the plus-strand DNA in hepatitis B virus. *J Virol*, 2011. **85**(22): p. 11916–27.
32. Bruss, V. and D. Ganem, *The role of envelope proteins in hepatitis* B virus assembly. *Proc Natl Acad Sci U S A*, 1991. **88**(3): p. 1059–63.
33. Ueda, K., T. Tsurimoto, and K. Matsubara, Three envelope proteins of hepatitis B virus: large S, middle S, and major S proteins needed for the formation of Dane particles. *J Virol*, 1991. **65**(7): p. 3521–9.
34. Kann, M. and W. Gerlich, *Chapter 10: Structure and Molecular Virology. Viral Hepatitis.* 3rd ed. Oxford, UK: Blackwell Publishing Ltd, 2005.
35. Prange, R., Host factors involved in hepatitis B virus maturation, assembly, and egress. *Med Microbiol Immunol*, 2012. **201**(4): p. 449–61.
36. Seeger, C. and W.S. Mason, Hepatitis B virus biology. *Microbiol Mol Biol Rev*, 2000. **64**(1): p. 51–68.
37. Cattaneo, R., H. Will, and H. Schaller, Hepatitis B virus transcription in the infected liver. *EMBO J*, 1984. **3**(9): p. 2191–6.
38. Nassal, M., M. Junker-Niepmann, and H. Schaller, Translational inactivation of RNA function: discrimination against a subset of genomic transcripts during HBV nucleocapsid assembly. *Cell*, 1990. **63**(6): p. 1357–63.
39. Standring, D.N., et al., A signal peptide encoded within the precore region of hepatitis B virus directs the secretion of a heterogeneous population of e antigens in Xenopus oocytes. *Proc Natl Acad Sci U S A*, 1988. **85**(22): p. 8405–9.
40. Wang, J., A.S. Lee, and J.H. Ou, Proteolytic conversion of hepatitis B virus e antigen precursor to end product occurs in a postendoplasmic reticulum compartment. *J Virol*, 1991. **65**(9): p. 5080–3.
41. Kimura, T., et al., Hepatitis B virus DNA-negative dane particles lack core protein but contain a 22-kDa precore protein without C-terminal arginine-rich domain. *J Biol Chem*, 2005. **280**(23): p. 21713–9.
42. Milich, D.R., et al., The secreted hepatitis B precore antigen can modulate the immune response to the nucleocapsid: a mechanism for persistence. *J Immunol*, 1998. **160**(4): p. 2013–21.
43. Crowther, R.A., et al., Three-dimensional structure of hepatitis B virus core particles determined by electron cryomicroscopy. *Cell*, 1994. **77**(6): p. 943–50.
44. Bruss, V., et al., Post-translational alterations in transmembrane topology of the hepatitis B virus large envelope protein. *EMBO J*, 1994. **13**(10): p. 2273–9.
45. Sivasudhan, E., et al., Hepatitis B viral protein HBx and the molecular mechanisms modulating the hallmarks of hepatocellular carcinoma: A comprehensive review. *Cells*, 2022. **11**(4).
46. Martin-Vilchez, S., et al., The molecular and pathophysiological implications of hepatitis B X antigen in chronic hepatitis B virus infection. *Rev Med Virol*, 2011. **21**(5): p. 315–29.

47. Henkler, F., et al., Intracellular localization of the hepatitis B virus HBx protein. *J Gen Virol*, 2001. **82**(Pt 4): p. 871–82.
48. Bouchard, M.J. and R.J. Schneider, The enigmatic X gene of hepatitis B virus. *J Virol*, 2004. **78**(23): p. 12725–34.
49. Faktor, O. and Y. Shaul, The identification of hepatitis B virus X gene responsive elements reveals functional similarity of X and HTLV-I tax. *Oncogene*, 1990. **5**(6): p. 867–72.
50. Seifer, M., et al., In vitro tumorigenicity of hepatitis B virus DNA and HBx protein. *J Hepatol*, 1991. **13**(Suppl 4): p. S61–5.
51. Paterlini, P., et al., Selective accumulation of the X transcript of hepatitis B virus in patients negative for hepatitis B surface antigen with hepatocellular carcinoma. *Hepatology*, 1995. **21**(2): p. 313–21.
52. Beck, J. and M. Nassal, Hepatitis B virus replication. *World J Gastroenterol*, 2007. **13**(1): p. 48–64.
53. Park, S.G. and G. Jung, Human hepatitis B virus polymerase interacts with the molecular chaperonin Hsp60. *J Virol*, 2001. **75**(15): p. 6962–8.
54. Beck, J. and M. Nassal, Efficient Hsp90-independent in vitro activation by Hsc70 and Hsp40 of duck hepatitis B virus reverse transcriptase, an assumed Hsp90 client protein. *J Biol Chem*, 2003. **278**(38): p. 36128–38.
55. Junker-Niepmann, M., R. Bartenschlager, and H. Schaller, A short cis-acting sequence is required for hepatitis B virus pregenome encapsidation and sufficient for packaging of foreign RNA. *EMBO J*, 1990. **9**(10): p. 3389–96.
56. Beck, J. and M. Nassal, Formation of a functional hepatitis B virus replication initiation complex involves a major structural alteration in the RNA template. *Mol Cell Biol*, 1998. **18**(11): p. 6265–72.
57. Tavis, J.E., B. Massey, and Y. Gong, The duck hepatitis B virus polymerase is activated by its RNA packaging signal, epsilon. *J Virol*, 1998. **72**(7): p. 5789–96.
58. Tavis, J.E. and D. Ganem, Evidence for activation of the hepatitis B virus polymerase by binding of its RNA template. *J Virol*, 1996. **70**(9): p. 5741–50.
59. Flodell, S., et al., The apical stem-loop of the hepatitis B virus encapsidation signal folds into a stable tri-loop with two underlying pyrimidine bulges. *Nucleic Acids Res*, 2002. **30**(21): p. 4803–11.
60. Flodell, S., et al., Structure elucidation of the hepatitis B virus encapsidation signal by NMR on selectively labeled RNAs. *J Biomol Struct Dyn*, 2002. **19**(4): p. 627–36.
61. Staprans, S., D.D. Loeb, and D. Ganem, Mutations affecting hepadnavirus plus-strand DNA synthesis dissociate primer cleavage from translocation and reveal the origin of linear viral DNA. *J Virol*, 1991. **65**(3): p. 1255–62.
62. Liu, Z., et al., Distribution of hepatitis B virus genotypes and subgenotypes: a meta-analysis. *Medicine (Baltimore)*, 2021. **100**(50): p. e27941.
63. Buti, M., HBeAg-positive chronic hepatitis B: why do i treat my patients with Nucleos(t)ide analogs? *Liver Int*, 2014. **34**(Suppl 1): p. 108–11.
64. Qiu, L.P., L. Chen, and K.P. Chen, Antihepatitis B therapy: a review of current medications and novel small molecule inhibitors. *Fundam Clin Pharmacol*, 2014. **28**(4): p. 364–381. doi: 10.1111/fcp.12053. Epub 2013 Nov 18. PMID: 24118072.
65. Kim, Y.J., et al., Tenofovir rescue therapy for chronic hepatitis B patients after multiple treatment failures. *World J Gastroenterol*, 2012. **18**(47): p. 6996–7002.
66. Deng, L. and H. Tang, Hepatitis B virus drug resistance to current nucleos(t)ide analogs: mechanisms and mutation sites. *Hepatol Res*, 2011. **41**(11): p. 1017–24.
67. World Health, O., Hepatitis B vaccines: WHO position paper, July 2017 - Recommendations. *Vaccine*, 2019. **37**(2): p. 223–5.

16 Bacteriophages

Vishnu Kumar
Institute of Anatomy and Cell Biology, Unit of Reproductive
Biology, Justus-Liebig-University Giessen, Giessen, Germany

INTRODUCTION

As mentioned in the previous chapter, viruses are obligate parasites that infect a wide range of host cells, including plants, animals, fungi, protists, bacteria, and archaea. Bacteriophages, commonly known as phages, are viruses that specifically target and infect bacteria. Like all viruses, bacteriophages exhibit a high degree of species specificity concerning their hosts. Typically, they infect only a single bacterial species or sometimes even specific strains within a species. Phages have been found in environments where bacteria can grow (Ranveer, Dasriya et al. 2024). Phages regulate the bacterial population by eliminating approximately 40% of bacterial biomass daily (Czajkowski, Jackson et al. 2019). The unique properties of phages that infect bacteria make them key players in resolving the antibiotic resistance problem in modern medicine. Furthermore, bacteriophages can be used in the food industry to make food safer in specific stages of food production. However, bacteriophages also facilitate the conversion of harmless bacteria to pathogenic bacteria.

In this chapter, we will discuss the historical journey of phage discovery, its classification, and its structure. Furthermore, we will explore the phage life cycle and its diverse applications.

HISTORY OF PHAGES

The discovery of bacteriophages has a complex history. In 1896, British bacteriologist Ernest Hankin observed antibacterial activity against *Vibrio cholerae* in India's waters of the Ganges and Jumna rivers. He proposed the presence of an unidentified substance, heat-labile and capable of passing through fine porcelain filters, as responsible for limiting the spread of cholera epidemics (Hankin 1896). Subsequently, Russian bacteriologist Gamaleya observed a similar phenomenon while studying *Bacillus subtilis* in 1898 (Samsygina and Boni 1984), and other investigators made related observations (van Helvoort 1992). Despite these early observations, further exploration of bacteriophages was limited until almost 20 years later when Frederick Twort, a bacteriologist from England, revisited the subject. In 1915, Twort reported similar phenomena and hypothesized the involvement of a virus, among other possibilities (Twort 1915). However, due to the onset of World War I and financial constraints (Twort 1915, Summers 2000), Twort did not pursue this discovery further. Independently, in 1917, French-Canadian microbiologist Felix d'Herelle, working at the Institut Pasteur in Paris, proposed that clear spots observed on petri dishes cultured with dysentery bacteria and nutrient agar were caused by an invisible microbe—a virus parasitic on bacteria. d'Herelle coined the term "bacteriophage," derived from *bacteria* and *phagein* ("to eat or devour," in Greek), to signify eaters of bacteria. In the 1920–1930s, phages were used to treat bacterial infections in Europe and the Soviet Union, but after the discovery of penicillin, the use of bacteriophages as antimicrobial agents decreased. They are seen as a possible therapy against multidrug-resistant strains of many bacteria.

CLASSIFICATION OF BACTERIOPHAGES

Bacteriophages naturally exhibit extensive diversity in their morphological and genomic composition. The Bacterial and Archaeal Viruses Subcommittee (BAVS) within the International

Committee on Taxonomy of Viruses (ICTV) is responsible for phage taxonomy, classifying them based on various properties such as genome composition (ss/ds, DNA, or RNA), morphology, capsid structure, and host range (Dion, Oechslin et al. 2020). Previously, bacteriophages were categorized into different families, including *Leviviridae*, *Microviridae*, *Inoviridae*, *Cystoviridae*, *Tectiviridae*, *Corticoviridae*, *Siphoviridae*, *Podoviridae*, and *Myoviridae* (Sanz-Gaitero, Seoane-Blanco et al. 2021). Table 16.1 lists these families along with their genome type, morphology, and examples. Notably, *Siphoviridae*, *Podoviridae*, and *Myoviridae* belong to the *Caudovirales* order, encompassing tailed bacteriophages with icosahedral or prolate capsids and linear double-stranded DNA genomes. However, in a recent update in August 2022, the ICTV revised the phage classification system, removing major families such as *Siphoviridae*, *Podoviridae*, and *Myoviridae* (Turner, Shkoporov et al. 2023). Additionally, the order *Caudovirales* was replaced by the class *Caudoviricetes*, grouping all tailed bacterial and archaeal viruses with icosahedral capsids and double-stranded DNA genomes. Most isolated phages to date are tailed and possess double-stranded DNA genomes (Ackermann 2007).

Morphologically, phages can exist in various forms, including polyhedral, filamentous, tailed, or pleomorphic particles (Sanz-Gaitero, Seoane-Blanco et al. 2021). Polyhedral phages share an icosahedral capsid shape but differ in genome types, such as double-stranded DNA (*Tectiviridae*, *Corticoviridae*), single-stranded DNA (*Microviridae*), single-stranded RNA (*Leviviridae*), and double-stranded RNA (*Cystoviridae*). Pleomorphic viruses exhibit diverse structures, including a lipoprotein envelope in the *Plasmaviridae* family of bacteriophages, lemon-shaped capsids (*Fuselloviridae*), droplet-like capsids (*Guttaviridae*), or bottle-shaped capsids (*Ampullaviridae*) of archaeal viruses (Krupovic, Prangishvili et al. 2011, Casey, Coffey et al. 2021). Filamentous bacteriophages, like those in the *Inoviridae* family, are composed of thousands of identical coat proteins arranged in a helical sheath around a central core containing single-stranded DNA. Interestingly, filamentous phages replicate and exit host cells without causing cell lysis. Bacteriophages exhibit significant variation in genome sizes, ranging from the small *Leuconostoc* phage L5 genome (2,435 bp) to the large *Bacillus megaterium* phage G genome (approximately 498,000 bp) (Casey, Coffey et al. 2021).

Family	Genome	Morphology	Examples
Leviviridae	Linear ssRNA (+)	Icosahedral capsid	MS2, Qβ, F1
Microviridae	Circular ssDNA (+)	Icosahedral capsid	φX174,G4, α3
Inoviridae	Circular ssDNA (+)	Rod-shaped filamentous, helical	M13
Cystoviridae	dsRNA (3 segments)	Double capsid structure (nucleocapsid) with an external lipid membrane	φ6
Tectiviridae	dsDNA	• Icosahedral capsid that is decorated with spikes at the fivefold vertices • Capsid encloses an internal host-derived membrane • Do not have tails	PRD1
Corticoviridae	Circular dsDNA	• Icosahedral capsid • Contains an internal lipid membrane	PM2
Plasmaviridae	Circular supercoiled dsDNA	Pleomorphic, membrane-enveloped viruses that infect bacteria without a cell wall	L2
Podoviridae	Linear dsDNA	• Icosahedral capsid • Short, noncontractile tail	T7, T3, φ29, P22
Myoviridae	Linear dsDNA	• Icosahedral capsid • Long contractile tail	T4, Mu, P1, P2
Siphoviridae	Linear dsDNA	• Icosahedral capsid • Long, flexible tail	λ, T5, HK97, N15

Bacteriophages

STRUCTURE OF BACTERIOPHAGES

Generally, phages possess DNA or RNA genomes stored within highly symmetric protein capsids for protection. These capsids can be icosahedral, helical, or other shapes and may be covered with an outer lipid membrane (e.g., *Cystoviridae*) or have a lipid membrane covering the capsid (e.g., *Tectiviridae*, *Corticoviridae*). In addition to genome protection, phages require apparatuses for host cell recognition. Simple phages like *Leviviridae* have a single minor capsid protein for this purpose, while more complex phages like *Inoviridae* and *Tectiviridae* dedicate multiple proteins to host cell recognition. *Caudovirales* possess highly efficient tail protein complexes for DNA transfer. Moreover, complex phage particles may contain proteins for environmental sensing, binding to suitable matrices, and other functions.

Extensive research has been conducted on the T4 bacteriophage. The structure of T4 viruses is as follows.

The bacteriophage T4 virus belongs to the *Myoviridae* family that infects *Escherichia coli* and carries 172 kb of linear dsDNA packed within its elongated icosahedral head (Baschong, Baschong-Prescianotto et al. 1991). The virus structure comprises four parts: head (capsid), neck, tail, and baseplate, as shown in Figure 16.1. The head is composed of over 3,000 polypeptide chains encompassing at least 12 different kinds of proteins, including the three essential proteins of the capsid: gene product (gp) 23, gp24, and gp20. The icosahedral caps and the midsection are formed by 160 hexamers of the gp23. The 12 pentagonal vertices of the icosahedron are occupied by 11 pentamers of gp24 and one dodecamer of gp20, which substitutes the 12th pentamer of gp24 at the portal vertex (Chen, Sun et al. 2017). The shell is adorned on the outside with gp hoc and gp soc, denoting highly antigenic outer capsid protein and small outer capsid protein, respectively (Ishii, Yamaguchi et al. 1978; Steven, Greenstone et al. 1992).

The bacteriophage T4 tail consists of two concentric protein cylinders: the inner cylinder, called the tail tube; and the outer cylinder, called the tail sheath. The tail tube is constructed from 144 copies of gp19 and features a 40 Å-diameter channel for DNA passage from the head to the infected cell. The tail sheath has a 90 Å-diameter and a width of approximately 210 Å, composed of 144 copies of gp18 (King and Mykolajewycz 1973; Moody and Makowski 1981; Lepault and Leonard 1985). Notably, the length of the tail tube remains constant during sheath contraction, resulting in almost half of the tube protruding from the contracted tail sheath and baseplate. The proximal end of the tail tube and sheath is halted by gp3 and gp15, respectively. Recognition of gp15 by gp13 and/or gp14 facilitates the joining of the head and the tail.

The neck of the bacteriophage T4 is located between the portal vertex gp20 and the connector protein gp15 and contains the gp13–gp14 complex. There are 12 fibritin molecules attached to the neck of the T4 phage. Six of these molecules align approximately in one plane, forming the phage collar, while the remaining six stretch downward to form the whiskers (Fokine, Zhang et al. 2013). Both the collar and whiskers act as molecular chaperones, aiding in the attachment of the long-tail fibers to the phage. The collar maintains a consistent structure regardless of whether the tail is extended or contracted. In contrast, the whiskers undergo conformational changes during tail contraction due to steric hindrances between the whiskers and the contracted tail sheath.

The distal end of the tail is attached with a multiprotein hexagonal baseplate that has a long-tail fiber (LTF) and a short-tail fiber (STF) at each corner. The baseplate comprises approximately 150 subunits derived from at least 16 distinct gene products. These proteins organize into six independently assembled wedges, converging around the central hub facilitated by trimeric proteins (gp9) and (gp12). The hub itself likely comprises (gp5), (gp27), gp29, and potentially gp28. Structurally, the baseplate exhibits a dome-shaped configuration featuring a central spike generated by the membrane-puncturing mechanism composed of the gp5–gp27 complex (Leiman, Kanamaru et al. 2003). During infection, the baseplate undergoes a significant conformational change from a dome-shaped to a flat, star-shaped structure. The spike punctures the cell membrane, and the lysozyme domain of

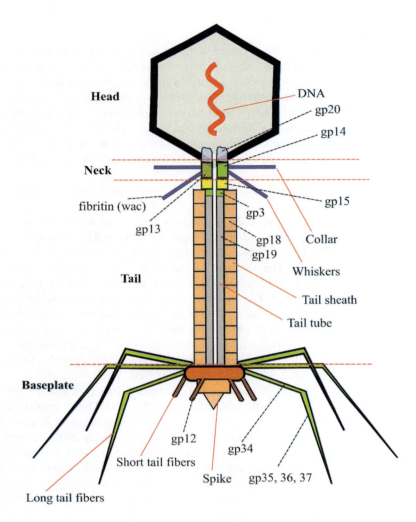

FIGURE 16.1 Structure of T4 bacteriophage. Dotted red line divided phage into four major parts: head, neck, tail, and baseplate. Red lines show the other components of the phage. Black dotted lines indicate the protein of respective parts.

gp5 digests the peptidoglycan in the *E. coli* periplasm. Additionally, the gp27 trimer forms a channel suitable for the passage of dsDNA and serves as an extension of the tail tube.

The LTFs function as adsorption devices and environmental sensors, linked to the baseplate via gp9 and gp7. Each fiber comprises rigid proximal and distal segments connected by a hinge region. The proximal part is formed by the gp34 trimer, while the distal part consists of the trimeric gp36 and gp37, along with the monomeric gp35. On the other hand, the STF consists of a single protein, gp12, forming a parallel, in-register homotrimer of 527 residues per subunit, which attaches to the baseplate via gp11, with the C termini of gp12 binding to the bacterial host cell.

LIFE CYCLE OF BACTERIOPHAGES

Like other viruses, phages carry out specific reactions to multiply and survive. They need to find a suitable host (bacterium), inject their genetic material into it, and use the bacterium's machinery to make many new phage particles. These phage particles are then assembled into new virions and released from the host cell through cell lysis to infect other cells. Phages have diverse life cycles, including lytic, lysogenic, and pseudolysogenic cycles (as shown in Figure 16.2). In the lytic cycle, phages immediately begin producing new viral progenies after infecting a host cell, which are then released through bacterial lysis. Conversely, in the lysogenic cycle, the phage genome exists as a prophage, either integrating into the host chromosome or remaining in a free plasmid-like state. This results in a long-term stable coexistence with the host, during which the phage persists in a dormant state that does not cause cell death or the production of phage particles. Upon induction, prophages in the lysogenic cycle can transition into the lytic cycle. Pseudolysogeny represents a non-classical phage life cycle where the phage genome exists in a non-replicative preprophage/episome stage due to specific conditions within the host cell, such as starvation (Feiner, Argov et al. 2015). However, when conditions improve, it can switch to either the lysogenic or lytic cycles. Here, we only elucidate the classical phage life cycles, i.e., the lytic and lysogenic cycles, because pseudolysogeny remains largely unknown.

Lytic Cycle

The lytic cycle leads to the lysis of host cells and the release of newly formed viral progeny. Phages that undergo the lytic cycle are referred to as virulent bacteriophages or lytic phages. An example of a virulent bacteriophage is T4. The lytic cycle consists of five main steps: attachment/adsorption, penetration, synthesis, assembly, and release (shown in Figure 16.2) (Willey, Sherwood et al. 2014).

Here, we are describing the lytic cycle of the T4 virulent phage. The cycle begins with the attachment of the bacteriophage to the host cell wall receptor. In the case of the T4 bacteriophage, the attachment process initiates when LTF contacts the lipopolysaccharide or certain proteins in the outer membrane of its *Escherichia coli* host. As more LTFs make contact with the surface, STFs extend downward, and the baseplate settles on the cell surface. Once the baseplate is settled, the tail sheath contracts, pushing the central tube through the outer membrane. Baseplate protein gp5 is released into the periplasm and penetrates the peptidoglycan through its lysozyme activity. Protein gp29, which links the baseplate to the sheath, moves into the periplasm. It then travels to the plasma membrane to form a channel through which the DNA is ejected.

After attachment, the phage genome is injected into the host cells (penetration). Once inside the cytoplasm, *Escherichia coli* RNA polymerase begins synthesizing T4 early mRNA. One of these early T4 genes encodes a protein that binds to the host enzyme RNaseE, directing it to degrade host mRNA and free host ribonucleotides and ribosomes for the transcription and translation of T4 genes. The virus-encoded DNA-dependent DNA polymerase initiates viral DNA replication bidirectionally at several origins of replication. This replication is followed by the synthesis of late mRNAs, encoding proteins vital for later stages.

The T4 genome also encodes the enzymes that synthesize hydroxymethyl cytosine (HMC), a modified nucleotide that replaces cytosine in T4 DNA. After T4 DNA synthesis, glucose is added to HMC residues through glycosylation. Glycosylated HMC residues protect T4 DNA from the restriction enzymes (endonuclease) of *E. coli*. The resulting T4 DNA is involved in forming a concatemer composed of multiple genome units linked together in the same orientation. During assembly, concatemers are cleaved such that the genome packaged in the capsid is slightly longer than the T4 gene set. Consequently, each progeny virus carries a genome unit beginning with a different gene and ending with the same set of genes.

Filling the head portion of the virion with the T4 genome is crucial and achieved by a complex of proteins known as the packasome. This includes terminase, which cuts the concatemers formed during T4 genome replication and coordinates DNA insertion into the T4 head. Terminase threads the T4 genome through a portal using ATP hydrolysis energy. The packasome is proposed to move the DNA by transitioning from B-form DNA to A-form DNA and back to B-form DNA, compressing the helix like a spring. When the phage head is filled with a DNA molecule about 3% longer than the length of one set of T4 genes, terminase makes a second cut, completing the packaging process

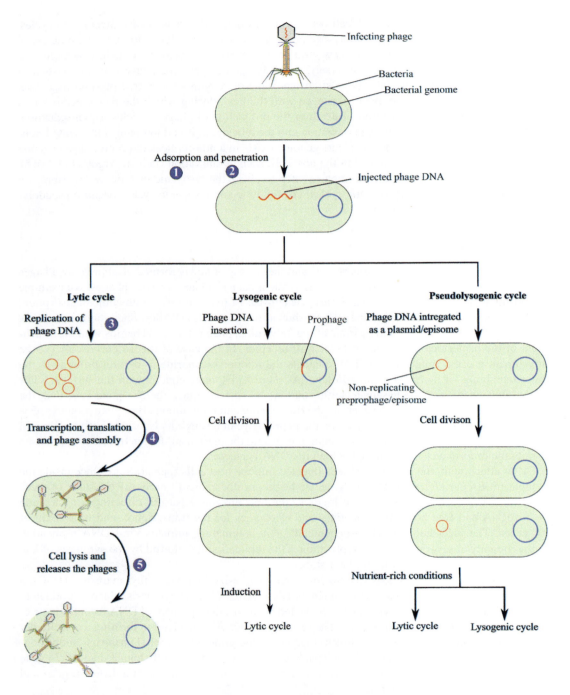

FIGURE 16.2 Schematic illustration showing the life cycle of bacteriophages.

for that head. Terminase then exits the head, and other viral proteins bind at the portal, sealing the head and preparing it for tail and tail fiber addition.

The assembly process results in the formation of numerous intact phage particles within the cell. When the virus particles reach around 150, T4 lyses *Escherichia coli* with the help of two proteins. One protein, holin, creates a pore in the plasma membrane, and the other, endolysin (also known as

Bacteriophages 125

T4 lysozyme), degrades peptidoglycan in the host's cell wall. With the assistance of these proteins, virions are released to infect other cells and initiate a new cycle.

LYSOGENIC CYCLE

In the lysogenic cycle, the phage genome integrates into the host chromosome, forming what is known as a prophage, and replicates alongside the host cell. The prophage remains in a dormant state, without causing cell death or the production of phage particles. Prophages can exist in two forms: integrated with host chromosomes, as found in bacteriophage lambda, or as extrachromosomal plasmids known as episomes, present in bacteriophage P1. Phages undergoing the lysogenic cycle are termed lysogenic viruses or temperate phages. A bacterium containing a prophage is referred to as a lysogen.

Temperate bacteriophages can enter either the lytic or lysogenic cycle upon infecting a host cell. The decision of a temperate phage to follow the lytic or lysogenic cycle depends on whether staying dormant (lysogeny) is more advantageous than using the host's resources right away to produce new phage particles. For example, bacteriophage lambda makes this decision based on signals derived from the cAMP and ppGpp alarmones levels (Łoś, Zielińska et al. 2021). When these alarmones are low, indicating poor resources or environmental conditions, bacteriophage lambda prefers the prophage state and follows the lysogenic cycle. Another factor influencing this decision is the density of bacteriophages in the environment. High local phage densities suggest that the environment is already saturated with phage, and the number of uninfected host cells may be very low. In such situations, lysogenizing the cell and reserving its resources for the potential production of progeny phages is the most rational choice.

Example of Lysogenic Cycle

The bacteriophage lambda (λ) DNA genome is a linear molecule with 12-nucleotide long cohesive ends (single-stranded stretches) that are complementary to each other and can base-pair. Similar to the T4 phage, it also uses its tail to inject its DNA into the host cell. Once inside, the cohesive ends of the linear genome pair up and circularize, and the host cell's DNA ligase seals the breaks in the strands. Transcription of the λ genome is carried out by the host cell's DNA-dependent RNA polymerase. This process begins at the phage promoters P_L (promoter left) and P_R (promoter right), transcribing the N and cro genes, respectively, as shown in Figure 16.3. The function of cro protein is to induce the lytic cycle. The product of the N gene acts as an anti-terminator factor, binds to RNA polymerase, and prevents transcriptional termination. Due to the action of the N protein, cII and cIII genes also get transcribed by P_R and P_L, respectively. The protein cII is an activator that is pivotal in determining if λ phage will establish lysogeny or follow a lytic pathway. Its high concentrations initiate the lysogenic cycle. However, the cII protein is susceptible to degradation by the host protease FtsH. The cIII protein binds to cII to prevent this degradation, thereby protecting it from proteolytic breakdown. The high cII protein levels increase the transcription of the int gene from P_I promoter, which encodes the enzyme integrase. Integrase catalyzes the integration of the λ genome into the host cell's chromosome, thus establishing lysogeny. This integration occurs via reciprocal recombination between the phage attachment site (attP) and the bacterial attachment site (attB), mediated by host-encoded proteins like integration host factor (IHF) and factor for inversion stimulation (Fis). The resulting prophage is integrated into the host chromosome at attB and is located between galactose (gal) and biotin (bio) operons (Willey, Sherwood et al. 2014). Once integrated, the prophage can persist indefinitely, undergoing replication alongside the bacterial genome.

Under certain conditions, such as ultraviolet light or chemical mutagens that damage DNA, the lysogenic state of prophage transit to the lytic cycle and this process is known as induction. In general, the RecA protein of the host is involved in the recombination and DNA repair. Due to the Uv DNA damage, activated RecA protein interacts with λ repressor, causing the repressor to cleave itself. As the level of the repressor becomes so low, the transcription of the xis, int, and cro genes

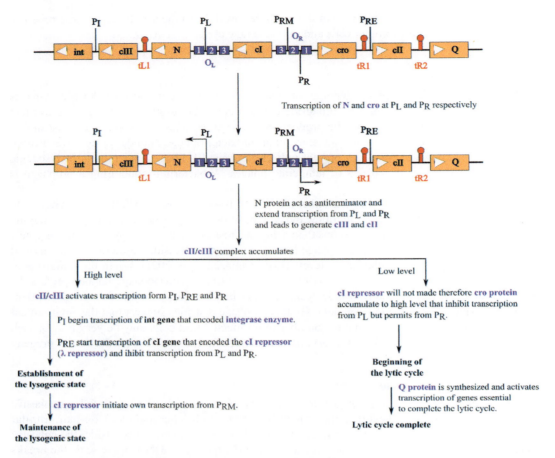

FIGURE 16.3 Schematic diagram of decision-making process for establishing lysogeny or the lytic pathway of lambda phage. In red color, tL1, tR1, and tR2 represent the terminators.

begins. The xis gene encodes the protein excisionase. Excisionase binds integrase, causing it to reverse the integration process, and the prophage is freed from the host chromosome. As λ repressor levels continue to decline, the cro protein levels increase. When the synthesis of λ repressor is completely blocked, protein Q levels increase, and the lytic cycle proceeds to completion.

Lysogeny is maintained by the cI repressor protein, also known as λ repressor. It is encoded by the cI gene and represses transcription of all bacteriophage genes (except its own). The cII protein facilitates the activation of cI transcription. Along with cIII, cII proteins assist in the generation of the cI repressor via the P_{RE} (promoter for repressor establishment) promoter. The λ repressor binds to operator sites adjacent to the P_R and P_L promoters. Both P_R and P_L possess three operator regions, denoted as O_{R1}, O_{R2}, O_{R3}, and O_{L1}, O_{L2}, O_{L3}, respectively. Among these, λ repressor exhibits the highest affinity for O_{R1}, followed by O_{R2} and O_{R3} (Atsumi and Little 2006). This arrangement dictates that $O_{R1} > O_{R2} > O_{R3}$ and $O_{L1} > O_{L2} > O_{L3}$ in terms of affinity. O_{R2} and O_{R3} also serve as operators for the P_{RM} (promoter of λ-repressor maintenance) promoter, situated adjacent to P_R. Binding to the O_{R1} region inhibits P_R, preventing transcription initiation and consequently suppressing the production of the cro protein. Conversely, when cI binds to the O_{R2} region, it stimulates the expression of cI with the help of P_{RM}. Notably, if P_R is blocked, P_{RM} initiates expression, although they cannot both be active simultaneously. Furthermore, binding to the O_{R3} region results in the repression of cI synthesis. Additionally, the cI repressor also acts upon P_L in the lysogenic state. In conclusion,

Bacteriophages

λ repressor maintains the lysogenic cycle by blocking the P_R and P_L while activating P_{RM}, initiating transcription by RNA polymerase and enabling the continuous synthesis of cI mRNA.

The cro protein, essential in the lytic cycle, binds to the same three sites at O_R and O_L as cI but with reversed affinities ($O_{R3} > O_{R2} = O_{R1}$, $OL3 > O_{L2} = O_{L1}$). When cro binds to O_{R3}, it blocks the transcription from P_{RM}, suppressing the expression of the cI repressor and thus halting the lysogenic cycle. Additionally, it inhibits transcription of the cIII and cI genes, reducing the levels of cII and λ repressor. However, it enhances its own production, along with another regulatory protein called Q. Once Q accumulates to a sufficient level, it activates the transcription of genes essential for the lytic cycle. cro also binds to three sites in the O_L region, though this repression is likely only partial. At higher concentrations of cro, it binds to O_{R2}/O_{R1} and inhibits transcription from P_R.

FACTORS THAT INFLUENCE THE CHOICE BETWEEN LYTIC AND LYSOGENIC CYCLE

The cII protein plays a crucial role in determining whether the cI repressor directs the cell toward the lysogenic or lytic cycle. Various factors influence the quantity of cII protein, including cell physiology, environmental conditions, and the number of phage particles infecting the cell. During starvation, the production of cyclic AMP is increased, which inhibits FtsH activity, leading to increased stability of cII. Consequently, cII protein accumulates more rapidly than cro protein, leading to the activation of the P_{RE} promoter to the formation of the cI repressor. Elevated levels of the cI repressor suppress the expression of the cro protein, thereby maintaining the lysogenic cycle. However, under growth conditions, high FtsH activity results in the degradation of cII, leading to reduced levels of the cI repressor and increased expression of the cro protein, which switches off the lysogenic cycle and initiates the lytic cycle (see Figure 16.4A).

Temperature also significantly affects the quantity of cII protein. At lower temperatures, the multimerization of cII protein (in its active tetrameric form) is more efficient. Additionally, FtsH-dependent degradation of cII tetramers is less effective than that of monomers. Moreover, the stability of the cII protein is influenced by the phage-encoded cIII protein, which is also a substrate for FtsH protease. Expression of the cIII gene is regulated by the P_L promoter, which exhibits higher activity at lower temperatures. Furthermore, the transcript for cIII can form two alternative structures, with only one capable of binding to the 30S ribosomal subunit and initiating translation. The proportion of these structures depends on temperature, with a higher proportion of transcripts able to bind to the ribosome at lower temperatures. In summary, low temperatures promote the lysogenic pathway, whereas high temperatures promote the lytic pathway (see Figure 16.4B).

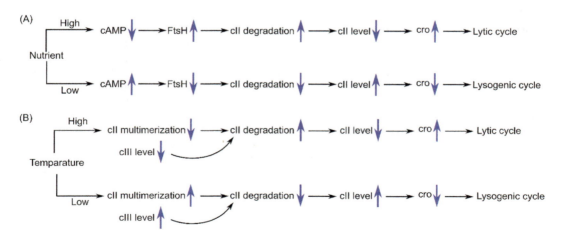

FIGURE 16.4 Schematic illustration showing the influence of (A) nutrients and (B) temperature on the decision-making process between the lytic and lysogenic cycles.

APPLICATION

Bacteriophages are host and strain-specific viruses that infect and kill bacteria without harming human and animal cells (Principi, Silvestri et al. 2019). Due to this unique ability, bacteriophages hold promise in various fields where bacterial infections are a concern.

As we know, food consumption is a common source of infectious diseases, serving as a vehicle for pathogen transmission, along with water and air. To address this issue, various strategies including sanitization by using physical (thermal, irradiation) or chemical approaches, cooked food, and antibiotics have been developed to reduce the presence of foodborne pathogens. Among these strategies, the use of bacteriophages has emerged as a promising approach. Bacteriophages can be employed at different stages to mitigate bacterial contamination, including in agricultural settings to combat diseases caused by pathogens such as *Pseudomonas* spp. and *Xanthomonas* spp. (Buttimer, McAuliffe et al. 2017). In the USA, products like AgriPhage, approved by the Food and Drug Administration (FDA) for agricultural use, have demonstrated effectiveness in controlling diseases such as pepper spot, speck, and tomato canker (Obradovic, Jones et al. 2005).

In food processing facilities, surface contamination poses a significant challenge. Bacteria can adhere to food contact surfaces, as well as in hospital settings, leading to potential contamination. Bacteriophages offer a solution as bio-sanitizers to clean contaminated surfaces. For example, *Listeria monocytogenes*, known for forming biofilms on various surfaces, can be targeted with phage products like ListShield™ and Listex™ (now called PhageGuardL) to reduce contamination levels (Gutiérrez, Rodríguez-Rubio et al. 2017). Similarly, healthcare settings are often contaminated with pathogens like *Staphylococcus* spp., including MRSA, *Enterobacteriaceae*, VRE, *Clostridium difficile*, and *Pseudomonas* spp. Phages can be utilized to reduce the concentration of these pathogens. For instance, phiIPLA-RODI and phiIPLA-C1C phages have been employed against mono- and dual-species staphylococcal biofilms (Gutierrez, Vandenheuvel et al. 2015).

Many common foodborne pathogenic bacteria can be found in the gastrointestinal tract of poultry, swine, and cattle, including *Campylobacter*, *Salmonella*, and enterohemorrhagic *Escherichia coli*. While these pathogens may not cause illness in animals, they can still pose risks to human health through contaminated food. With increasing concerns about antibiotic resistance, industries are exploring alternative methods, such as bacteriophage therapy, to reduce bacterial contamination in food. The following table provides a list of phages targeting foodborne pathogens (Hyla, Dusza et al. 2022).

Bacteria	Transmission	Phages	References
Campylobacter jejuni	Consumption of poultry meat and milk, exposure to contaminated water, swimming in contaminated water bodies, and contact with animals	CJ01, Φ7-izsam, Φ16-izsam, CP20, CP30A, 12673, P22, and 29C	(Goode, Allen et al. 2003; Richards, Connerton et al. 2019; Thung, Lee et al. 2019; D'Angelantonio, Scattolini et al. 2021)
Escherichia coli	Pork, poultry, contaminated ruminants (such as goats, deer, sheep, and elk), water, milk and dairy products, and direct animal contact	FM10, DP16, DP19, DW-EC, AZO145A, T4, and FAHEc1	(Hudson, Billington et al. 2015; Gwak, Choi et al. 2018; Mangieri, Picozzi et al. 2020; Wang, Hang et al. 2020; Dewanggana, Evangeline et al. 2022)

Bacteria	Transmission	Phages	References
Listeria monocytogenes	Fresh vegetables and fruit, dairy products that haven't undergone pasteurization (such as milk, cheese, and ice cream), poultry meat in various states (raw, cooked, and frozen), both raw and smoked fish, deli items, partially prepared foods, fast-food items, soil, sewage, water, decomposing plants, silage, and both wild and domesticated animals	FWLLm1 A511	(Bigot, Lee et al. 2011) (Guenther, Huwyler et al. 2009)
Pseudomonas spp.	Water, soil, and the digestive tract of humans and animals	UFJF_PfDIW6, UFJF_PfSW6, V523, V524, and JG003	(Kauppinen, Siponen et al. 2021; Nascimento, Sabino et al. 2022)
Salmonella spp.	Eggs, chicken, pork, turkey, duck, goose meat, cheese, milk, fruits, vegetables, soil, water, and contact with contaminated animals	LPSTLL, LPST94, LPST153, LYPSET, SE07, SJ2, BSPM4, BSP101, and BSP22A	(Hong, Schmidt et al. 2016; Thung, Krishanthi Jayarukshi Kumari Premarathne et al. 2017; Bai, Jeon et al. 2019; Islam, Zhou et al. 2019; Yan, Liang et al. 2020)
Shigella sp.	Coming into contact with the skin of an infected individual, ingesting contaminants through the oral cavity (via the fecal–oral route), consuming contaminated water and food, engaging in sexual contact, swimming in polluted water, and transmitting via insects like houseflies	SSE1, SGF3, SGF2, SD-11, SF-A2, and SS-92	(Zhang, Wang et al. 2013; Lu, Liu et al. 2020)
Staphylococcus sp.	Transmission primarily occurs through direct contact; patients who have recently undergone surgery are at the highest risk	MDR, ME18, and ME126	(Ali Gharieb, Saad et al. 2020)
Vibrio parahaemolyticus	Exposure to contaminated fruits, seafood, and water	PVP1 and PVP2	(Ren, Li et al. 2019)
Yersinia	Milk, water, raw vegetables, fruits, pork and pork offal	fHe-Yen3-01 fHe-Yen9-01, fHe-Yen9-02, and fHe-Yen9-03	(Jun, Park et al. 2018)

PHAGES FOR DIAGNOSIS

Phages grow and replicate in live bacteria and can be used for detection methods to enable the rapid and specific identification of viable bacterial cells. Broadly, methods for pathogen detection using phages can be divided into the following categories: phage display, reporter phage, phage amplification, and phage capture.

In the phage display system, an exogenous protein is expressed (displayed) on the surface of the phage by inserting an exogenous DNA fragment into the phage genome (in the phage coat protein-encoding gene region). This modified phage can be used in the development of phage ELISA for

130   Textbook of General Virology

antibody detection, biosensors for rapid pathogen identification, and peptide arrays for comprehensive biomarker screening.

In the reporter phage, the phage is genetically modified to express reporter proteins like luciferases or fluorescent proteins under the strong endogenous promoter (e.g., structural genes) to produce an intense and amplified signal when the phage infects viable hosts.

In the phage amplification assays, the production of progeny phage or the death of the bacterial host is used as a detection signal. The increased phage numbers indicate the presence of a susceptible host within the sample. The change in phage titer can be determined by traditional plaque assays using suitable indicator strains, by physical detection of the phage particles using ELISA-based assays, and by measuring the increase in phage nucleic acid content. The sensitivity of these assays can be improved by capturing and enriching the progeny phage particles using lateral flow assays or (magnetic) bead-based enrichment. In addition to progeny phage detection, cell lysis releases an abundance of cellular content that can be used as diagnostic markers. For instance, ATP is easily detectable after release from cells using a bioluminescence reaction with firefly luciferase.

Due to the inherent specificity of phages for their hosts (host-specific phage proteins), phages or phage components are also being used as capture elements to capture the bacteria. The cell wall-binding domains, or immobilized phage/tail spike, can be utilized to capture and isolate target organisms and can be detected further by culture, ELISA, or qPCR.

The following table summarizes the utilization of phages for diagnostics (Meile, Kilcher et al. 2020; Paczesny, Richter et al. 2020; Jo, Kwon et al. 2023)

Phages at the Surface

Phage	Target	Method	References
C4-22 phage	*Salmonella enterica*	Magnetoelastic	(Chen, Horikawa et al. 2017)
Gamma phages	*Bacillus thuringiensis*	SERS	(Lai, Almaviva et al. 2017)
M13 phage	Dengue 3 and 4 viruses	ELISA, immunofluorescence assay	(Cabezas, Rojas et al. 2009)
M13 phage for specific binding peptide screening	Ovarian cancer	SPECT/CT	(Soendergaard, Newton-Northup et al. 2014)
PaP1 phage	*Pseudomonas aeruginosa*	Electrochemiluminescence	(Yue, He et al. 2017)
Phage 12600	Methicillin-resistant *Staphylococcusaureus*	Magnetoelastic	(Hiremath, Chin et al. 2017)
T2 phage	*E. coli* B ATCC 11303	Electrochemical impedance spectroscopy	(Zhou, Marar et al. 2017)
T4 phage	*E.coli* BL21 DE3	Microscopic	(Richter, Matuła et al. 2016)
T4 phage	*E. coli* B	Differential pulse voltammetry	(Xu, Zhao et al. 2020)
T7 phage	Lung cancer	Protein chip	(Lee, Lee et al. 2016)

Phage Amplification-Based Detection Assays

Phage	Target	Method	References
A511	*Listeria monocytogenes*	Immunoassay, SERS-LFI	(Stambach, Carr et al. 2015)
CGG4-1	*Salmonella* Newport	qRT-PCR	(Anany, Brovko et al. 2018)
D29	*Mycobacterium avium* subsp. paratuberculosis (MAP)	Plaques, PCR	(Swift, Huxley et al. 2016)
D29	*Mycobacterium bovis*	DNA amplification	(Swift, Convery et al. 2016)
DN1, UP2, UP5	*S. Typhi* and *S. Paratyphi*	Optical, colorimetry	(Vaidya, Ravindranath et al. 2020)
MS2	*E. coli*	Immunoassay	(Mido, Schaffer et al. 2018)

PA phage	*Pseudomonas aeruginosa*	Plaques	(Ben Said, Ben Saad et al. 2021)
PAP1 phage	*Pseudomonas aeruginosa*	Luminescence	(He, Wang et al. 2017)
Phage 10	*S. Typhimurium*	Optical, absorbance	(Tamariz, Guevara et al. 2018)
Phage gamma	*Bacillus antracis*	qRT-PCR (RNA)	(Malagon, Estrella et al. 2020)
Phage K	*Staphylococcus* spp.	Bioluminescence	(Suster, Podgornik et al. 2017)
Phage K	*Staphylococcus aureus*	MALDI-MS	(Rees and Barr 2017)
Phage K	*Staphylococcus aureus*	qRT-PCR (RNA)	(Malagon, Estrella et al. 2020)
rV5, AG2A	*E. coli*	qRT-PCR	(Anany, Brovko et al. 2018)
ST560Ø	*Salmonella typhi*	Plaques	(Ben Said, Ben Saad et al. 2019)
T7	*E. coli*	Fluorescence microscopy	(Yang, Wisuthiphaet et al. 2020)
T7	*E. coli*	Colorimetry	(Chen, Jackson et al. 2016)
T7 phage	*E. coli* BL21	Fluorescence	(Tilton, Das et al. 2019)
Tb, Fz, Wb, S708, Bk	*Brucella abortus*	qRT-PCR	(Sergueev, Filippov et al. 2017)

Reporter Phage-Based Detection Assays

Phage	Target	Method	References
A511::nluc	*Listeria* spp.	Bioluminescence	(Meile, Sarbach et al. 2020)
NRGp2 (T7)	*E. coli*	Colorimetry	(Hinkley, Singh et al. 2018)
NRGp5 (T7)	*E. coli*	Bioluminescence	(Hinkley, Garing et al. 2020)
NRGp7 (T7)	*E. coli*	Electrochemistry	(Wang, Hinkley et al. 2019)
phiV10lux	*E. coli*	Bioluminescence	(Kim, Kim et al. 2017)
PP01ccp	*E. coli*	Colorimetry	(Hoang and Nhung 2018)
T7ALP	*E. coli*	Colorimetry	(Alcaine, Law et al. 2016)
T7LacZ	*E. coli*	Colorimetry	(Chen, Picard et al. 2017)
T7LacZ	*E. coli*	Electrochemistry	(Wang, Chen et al. 2017)
T7MBP	*E. coli*	Fluorescence	(Alcaine, Law et al. 2016)
Wβ::luxAB-2	*Bacillus anthracis*	Bioluminescence	(Sharp, Molineux et al. 2016)
Y2	*Erwinia amylovora*	Bioluminescence	(Born, Fieseler et al. 2017)
Φ2GFP10 (TM4-derived)	*Mycobacterium* spp.	Fluorescence	(O'Donnell, Larsen et al. 2019)
ΦV10	*E. coli*	Bioluminescence	(Zhang, Coronel-Aguilera et al. 2016)

Phage Capture

Phage	Target	Method	References
Salmonella phage S16 long-tail fiber protein	*S. enterica*	ELISA	(Filik, Szermer-Olearnik et al. 2022)
Yersinia phage φYeO3-12 tail fiber protein Gp17	*Y. enterocolitica* serotype O:3	ELISA	(Filik, Szermer-Olearnik et al. 2022)

CONCLUSION

Bacteriophages are the viruses that eat bacteria. These viruses are stain-specific and condition-specific, following the lytic and lysogenic cycles to replicate and survive in harsh conditions. Contaminated foods pose significant challenges for the food industry, leading to economic losses and health risks due to bacterial infections. Despite their primary application in phage therapy and food safety, bacteriophages are also being explored for their potential in diagnostics. However, further research is necessary to delve into the diversity of bacteriophages, understand their biology

REFERENCES

Ackermann, H. W. (2007). "5500 Phages examined in the electron microscope." *Arch Virol* **152**(2): 227–243.

Alcaine, S. D., K. Law, S. Ho, A. J. Kinchla, D. A. Sela and S. R. Nugen (2016). "Bioengineering bacteriophages to enhance the sensitivity of phage amplification-based paper fluidic detection of bacteria." *Biosens Bioelectron* **82**: 14–19.

Ali Gharieb, R. M., M. F. Saad, A. S. Mohamed and Y. H. Tartor (2020). "Characterization of two novel lytic bacteriophages for reducing biofilms of zoonotic multidrug-resistant Staphylococcus aureus and controlling their growth in milk." *Lwt* **124**: 109145.

Anany, H., L. Brovko, N. K. El Dougdoug, J. Sohar, H. Fenn, N. Alasiri, T. Jabrane, P. Mangin, M. Monsur Ali, B. Kannan, C. D. M. Filipe and M. W. Griffiths (2018). "Print to detect: A rapid and ultrasensitive phage-based dipstick assay for foodborne pathogens." *Anal Bioanal Chem* **410**(4): 1217–1230.

Atsumi, S. and J. W. Little (2006). "Role of the lytic repressor in prophage induction of phage lambda as analyzed by a module-replacement approach." *Proc Natl Acad Sci U S A* **103**(12): 4558–4563.

Bai, J., B. Jeon and S. Ryu (2019). "Effective inhibition of Salmonella Typhimurium in fresh produce by a phage cocktail targeting multiple host receptors." *Food Microbiol* **77**: 52–60.

Baschong, W., C. Baschong-Prescianotto, A. Engel, E. Kellenberger, A. Lustig, R. Reichelt, M. Zulauf and U. Aebi (1991). "Mass analysis of bacteriophage T4 proheads and mature heads by scanning transmission electron microscopy and hydrodynamic measurements." *J Struct Biol* **106**(2): 93–101.

Ben Said, M., M. Ben Saad, F. Achouri, L. Bousselmi and A. Ghrabi (2019). "Detection of active pathogenic bacteria under stress conditions using lytic and specific phage." *Water Sci Technol* **80**(2): 282–289.

Ben Said, M., M. Ben Saad, F. Achouri, L. Bousselmi and A. Ghrabi (2021). "The application of phage reactivation capacity to sens bacterial viability and activity after photocatalytic treatment." *Environ Technol* **42**(18): 2836–2844.

Bigot, B., W. J. Lee, L. McIntyre, T. Wilson, J. A. Hudson, C. Billington and J. A. Heinemann (2011). "Control of Listeria monocytogenes growth in a ready-to-eat poultry product using a bacteriophage." *Food Microbiol* **28**(8): 1448–1452.

Born, Y., L. Fieseler, V. Thony, N. Leimer, B. Duffy and M. J. Loessner (2017). "Engineering of bacteriophages Y2::dpoL1-C and Y2::luxAB for efficient control and rapid detection of the fire blight pathogen, Erwinia amylovora." *Appl Environ Microbiol* **83**(12).

Buttimer, C., O. McAuliffe, R. P. Ross, C. Hill, J. O'Mahony and A. Coffey (2017). "Bacteriophages and bacterial plant diseases." *Front Microbiol* **8**: 34.

Cabezas, S., G. Rojas, A. Pavon, L. Bernardo, Y. Castellanos, M. Alvarez, M. Pupo, G. Guillen and M. G. Guzman (2009). "Phage-displayed antibody fragments recognizing dengue 3 and dengue 4 viruses as tools for viral serotyping in sera from infected individuals." *Arch Virol* **154**(7): 1035–1045.

Casey, A., A. Coffey and O. McAuliffe (2021). "Genetics and genomics of bacteriophages." *Bacteriophages: Biology, Technology, Therapy* 193–218.

Chen, I. H., S. Horikawa, K. Bryant, R. Riggs, B. A. Chin and J. M. Barbaree (2017). "Bacterial assessment of phage magnetoelastic sensors for Salmonella enterica Typhimurium detection in chicken meat." *Food Control* **71**: 273–278.

Chen, J., A. A. Jackson, V. M. Rotello and S. R. Nugen (2016). "Colorimetric detection of escherichia coli based on the enzyme-induced metallization of gold nanorods." *Small* **12**(18): 2469–2475.

Chen, J., R. A. Picard, D. Wang and S. R. Nugen (2017). "Lyophilized engineered phages for escherichia coli detection in food matrices." *ACS Sens* **2**(11): 1573–1577.

Chen, Z., L. Sun, Z. Zhang, A. Fokine, V. Padilla-Sanchez, D. Hanein, W. Jiang, M. G. Rossmann and V. B. Rao (2017). "Cryo-EM structure of the bacteriophage T4 isometric head at 3.3-A resolution and its relevance to the assembly of icosahedral viruses." *Proc Natl Acad Sci U S A* **114**(39): E8184–E8193.

Czajkowski, R., R. W. Jackson and S. E. Lindow (2019). "Editorial: Environmental bacteriophages: From biological control applications to directed bacterial evolution." *Front Microbiol* **10**: 1830.

D'Angelantonio, D., S. Scattolini, A. Boni, D. Neri, G. Di Serafino, P. Connerton, I. Connerton, F. Pomilio, E. Di Giannatale, G. Migliorati and G. Aprea (2021). "Bacteriophage therapy to reduce colonization of campylobacter jejuni in broiler chickens before slaughter." *Viruses* **13**(8): 1428.

Dewanggana, M. N., C. Evangeline, M. D. Ketty, D. E. Waturangi, Yogiara and S. Magdalena (2022). "Isolation, characterization, molecular analysis and application of bacteriophage DW-EC to control Enterotoxigenic Escherichia coli on various foods." *Sci Rep* **12**(1): 495.

Dion, M. B., F. Oechslin and S. Moineau (2020). "Phage diversity, genomics and phylogeny." *Nat Rev Microbiol* **18**(3): 125–138.

Feiner, R., T. Argov, L. Rabinovich, N. Sigal, I. Borovok and A. A. Herskovits (2015). "A new perspective on lysogeny: prophages as active regulatory switches of bacteria." *Nat Rev Microbiol* **13**(10): 641–650.

Filik, K., B. Szermer-Olearnik, J. Niedziolka-Jonson, E. Rozniecka, J. Ciekot, A. Pyra, I. Matyjaszczyk, M. Skurnik and E. Brzozowska (2022). "phiYeO3-12 phage tail fiber Gp17 as a promising high specific tool for recognition of Yersinia enterocolitica pathogenic serotype O:3." *AMB Express* **12**(1): 1.

Fokine, A., Z. Zhang, S. Kanamaru, V. D. Bowman, A. A. Aksyuk, F. Arisaka, V. B. Rao and M. G. Rossmann (2013). "The molecular architecture of the bacteriophage T4 neck." *J Mol Biol* **425**(10): 1731–1744.

Goode, D., V. M. Allen and P. A. Barrow (2003). "Reduction of experimental Salmonella and Campylobacter contamination of chicken skin by application of lytic bacteriophages." *Appl Environ Microbiol* **69**(8): 5032–5036.

Guenther, S., D. Huwyler, S. Richard and M. J. Loessner (2009). "Virulent bacteriophage for efficient biocontrol of Listeria monocytogenes in ready-to-eat foods." *Appl Environ Microbiol* **75**(1): 93–100.

Gutiérrez, D., L. Rodríguez-Rubio, L. Fernández, B. Martínez, A. Rodríguez and P. García (2017). "Applicability of commercial phage-based products against Listeria monocytogenes for improvement of food safety in Spanish dry-cured ham and food contact surfaces." *Food Control* **73**: 1474–1482.

Gutierrez, D., D. Vandenheuvel, B. Martinez, A. Rodriguez, R. Lavigne and P. Garcia (2015). "Two phages, phiIPLA-RODI and phiIPLA-C1C, lyse mono- and dual-species staphylococcal biofilms." *Appl Environ Microbiol* **81**(10): 3336–3348.

Gwak, K. M., I. Y. Choi, J. Lee, J. H. Oh and M. K. Park (2018). "Isolation and characterization of a lytic and highly specific phage against yersinia enterocolitica as a novel biocontrol agent." *J Microbiol Biotechnol* **28**(11): 1946–1954.

Hankin, E. (1896). "L'action bactericide des eaux de la Jumna et du Gange sur le vibrion du cholera." *Ann Inst Pasteur* **10**: 511.

He, Y., M. Wang, E. Fan, H. Ouyang, H. Yue, X. Su, G. Liao, L. Wang, S. Lu and Z. Fu (2017). "Highly specific bacteriophage-affinity strategy for rapid separation and sensitive detection of viable pseudomonas aeruginosa." *Anal Chem* **89**(3): 1916–1921.

Hinkley, T. C., S. Garing, P. Jain, J. Williford, A. M. Le Ny, K. P. Nichols, J. E. Peters, J. N. Talbert and S. R. Nugen (2020). "A syringe-based biosensor to rapidly detect low levels of escherichia coli (ECOR13) in drinking water using engineered bacteriophages." *Sensors (Basel)* **20**(7): 1953.

Hinkley, T. C., S. Singh, S. Garing, A. M. Le Ny K. P. Nichols, J. E. Peters, J. N. Talbert and S. R. Nugen (2018). "A phage-based assay for the rapid, quantitative, and single CFU visualization of E. coli (ECOR #13) in drinking water." *Sci Rep* **8**(1): 14630.

Hiremath, N., B. A. Chin and M.-K. Park (2017). "Effect of competing foodborne pathogens on the selectivity and binding kinetics of a lytic phage for methicillin-Resistant Staphylococcus aureus Detection." *Journal of The Electrochemical Society* **164**(4): B142–B146.

Hoang, H. A. and N. T. T. Nhung (2018). "Development of a bacteriophage-based method for detection of escherichia coli O157:H7 in fresh vegetables." *Food Saf (Tokyo)* **6**(4): 143–150.

Hong, Y., K. Schmidt, D. Marks, S. Hatter, A. Marshall, L. Albino and P. Ebner (2016). "Treatment of salmonella-contaminated eggs and pork with a broad-spectrum, single bacteriophage: Assessment of efficacy and resistance development." *Foodborne Pathog Dis* **13**(12): 679–688.

Hudson, J. A., C. Billington, T. Wilson and S. L. On (2015). "Effect of phage and host concentration on the inactivation of Escherichia coli O157: H7 on cooked and raw beef." *Food Sci Technol Int* **21**(2): 104–109.

Hyla, K., I. Dusza and A. Skaradzinska (2022). "Recent advances in the application of bacteriophages against common foodborne pathogens." *Antibiotics (Basel)* **11**(11): 1536.

Ishii, T., Y. Yamaguchi and M. Yanagida (1978). "Binding of the structural protein soc to the head shell of bacteriophage T4." *J Mol Biol* **120**(4): 533–544.

Islam, M. S., Y. Zhou, L. Liang, I. Nime, K. Liu, T. Yan, X. Wang and J. Li (2019). "Application of a phage cocktail for control of salmonella in foods and reducing biofilms." *Viruses* **11**(9).

Jo, S. J., J. Kwon, S. G. Kim and S. J. Lee (2023). "The biotechnological application of bacteriophages: what to do and where to go in the middle of the post-antibiotic era." *Microorganisms* **11**(9): 2311.

Jun, J. W., S. C. Park, A. Wicklund and M. Skurnik (2018). "Bacteriophages reduce Yersinia enterocolitica contamination of food and kitchenware." *Int J Food Microbiol* **271**: 33–47.

Kauppinen, A., S. Siponen, T. Pitkanen, K. Holmfeldt, A. Pursiainen, E. Torvinen and I. T. Miettinen (2021). "Phage Biocontrol of Pseudomonas aeruginosa in Water." *Viruses* **13**(5): 928.

Kim, J., M. Kim, S. Kim and S. Ryu (2017). "Sensitive detection of viable Escherichia coli O157: H7 from foods using a luciferase-reporter phage phiV10lux." *Int J Food Microbiol* **254**: 11–17.

King, J. and N. Mykolajewycz (1973). "Bacteriophage T4 tail assembly: proteins of the sheath, core and baseplate." *J Mol Biol* **75**(2): 339–358.

Krupovic, M., D. Prangishvili, R. W. Hendrix and D. H. Bamford (2011). "Genomics of bacterial and archaeal viruses: dynamics within the prokaryotic virosphere." *Microbiol Mol Biol Rev* **75**(4): 610–635.

Lai, A., S. Almaviva, V. Spizzichino, D. Luciani, A. Palucci, S. Mengali, C. Marquette, O. Berthuy, B. Jankiewicz and L. Pierno (2017). "Bacillus spp. Cells captured selectively by phages and identified by surface enhanced raman spectroscopy technique." In Proceedings of Eurosensors 2017, volume 1, page **519**.

Lee, K. J., J. H. Lee, H. K. Chung, E. J. Ju, S. Y. Song, S. Y. Jeong and E. K. Choi (2016). "Application of peptide displaying phage as a novel diagnostic probe for human lung adenocarcinoma." *Amino Acids* **48**(4): 1079–1086.

Leiman, P. G., S. Kanamaru, V. V. Mesyanzhinov, F. Arisaka and M. G. Rossmann (2003). "Structure and morphogenesis of bacteriophage T4." *Cell Mol Life Sci* **60**(11): 2356–2370.

Lepault, J. and K. Leonard (1985). "Three-dimensional structure of unstained, frozen-hydrated extended tails of bacteriophage T4." *J Mol Biol* **182**(3): 431–441.

Łoś, J., S. Zielińska, A. Krajewska, Z. Michalina, A. Małachowska, K. Kwaśnicka and M. Łoś (2021). "Temperate phages, prophages, and lysogeny." Bacteriophages: biology, technology, therapy, pages119–150.

Lu, H., H. Liu, M. Lu, J. Wang, X. Liu and R. Liu (2020). "Isolation and characterization of a novel myovirus Infecting Shigella dysenteriae from the aeration tank water." *Appl Biochem Biotechnol* **192**(1): 120–131.

Malagon, F., L. A. Estrella, M. G. Stockelman, T. Hamilton, N. Teneza-Mora and B. Biswas (2020). "Phage-mediated molecular detection (PMMD): A novel rapid method for phage-specific bacterial detection." *Viruses* **12**(4).

Mangieri, N., C. Picozzi, R. Cocuzzi and R. Foschino (2020). "Evaluation of a potential bacteriophage cocktail for the control of shiga-toxin producing escherichia coli in food." *Front Microbiol* **11**: 1801.

Meile, S., S. Kilcher, M. J. Loessner and M. Dunne (2020). "Reporter phage-based detection of bacterial pathogens: Design guidelines and recent developments." *Viruses* **12**(9).

Meile, S., A. Sarbach, J. Du, M. Schuppler, C. Saez, M. J. Loessner and S. Kilcher (2020). "Engineered reporter phages for rapid bioluminescence-based detection and differentiation of viable listeria cells." *Appl Environ Microbiol* **86**(11):e00442–20.

Mido, T., E. M. Schaffer, R. W. Dorsey, S. Sozhamannan and E. R. Hofmann (2018). "Sensitive detection of live Escherichia coli by bacteriophage amplification-coupled immunoassay on the Luminex(R) MAGPIX instrument." *J Microbiol Methods* **152**: 143–147.

Moody, M. F. and L. Makowski (1981). "X-ray diffraction study of tail-tubes from bacteriophage T2L." *J Mol Biol* **150**(2): 217–244.

Nascimento, E. C. D., M. C. Sabino, L. D. R. Corguinha, B. N. Targino, C. C. Lange, C. L. O. Pinto, P. F. Pinto, P. M. P. Vidigal, A. S. Sant'Ana and H. M. Hungaro (2022). "Lytic bacteriophages UFJF_PfDIW6 and UFJF_PfSW6 prevent Pseudomonas fluorescens growth in vitro and the proteolytic-caused spoilage of raw milk during chilled storage." *Food Microbiol* **101**: 103892.

O'Donnell, M. R., M. H. Larsen, T. S. Brown, P. Jain, V. Munsamy, A. Wolf, L. Uccellini, F. Karim, T. de Oliveira, B. Mathema, W. R. Jacobs and A. Pym (2019). "Early detection of emergent extensively drug-resistant tuberculosis by flow cytometry-based phenotyping and whole-genome sequencing." *Antimicrob Agents Chemother* **63**(4): 10–1128.

Obradovic, A., J. B. Jones, M. T. Momol, S. M. Olson, L. E. Jackson, B. Balogh, K. Guven and F. B. Iriarte (2005). "Integration of biological control agents and systemic acquired resistance inducers against bacterial spot on tomato." *Plant Dis* **89**(7): 712–716.

Paczesny, J., L. Richter and R. Holyst (2020). "Recent progress in the detection of bacteria using bacteriophages: A review." *Viruses* **12**(8): 845.

Principi, N., E. Silvestri and S. Esposito (2019). "Advantages and limitations of bacteriophages for the treatment of bacterial infections." *Front Pharmacol* **10**: 513.

Ranveer, S. A., V. Dasriya, M. F. Ahmad, H. S. Dhillon, M. Samtiya, E. Shama, T. Anand, T. Dhewa, V. Chaudhary, P. Chaudhary, P. Behare, C. Ram, D. V. Puniya, G. D. Khedkar, A. Raposo, H. Han and A. K. Puniya (2024). "Positive and negative aspects of bacteriophages and their immense role in the food chain." *NPJ Sci Food* **8**(1): 1.

Rees, J. C. and J. R. Barr (2017). "Detection of methicillin-resistant Staphylococcus aureus using phage amplification combined with matrix-assisted laser desorption/ionization mass spectrometry." *Anal Bioanal Chem* **409**(5): 1379–1386.

Ren, H., Z. Li, Y. Xu, L. Wang and X. Li (2019). "Protective effectiveness of feeding phage cocktails in controlling Vibrio parahaemolyticus infection of sea cucumber Apostichopus japonicus." *Aquaculture* **503**: 322–329.

Richards, P. J., P. L. Connerton and I. F. Connerton (2019). "Phage biocontrol of campylobacter jejuni in chickens does not produce collateral effects on the gut microbiota." *Front Microbiol* **10**: 476.

Richter, Ł., K. Matuła, A. Leśniewski, K. Kwaśnicka, J. Łoś, M. Łoś, J. Paczesny and R. Hołyst (2016). "Ordering of bacteriophages in the electric field: Application for bacteria detection." *Sensors and Actuators B: Chemical* **224**: 233–240.

Samsygina, G. A. and E. G. Boni (1984). "Bacteriophages and phage therapy in pediatric practice." *Pediatriia* (4): 67–70.

Sanz-Gaitero, M., M. Seoane-Blanco and M. J. van Raaij (2021). "Structure and function of bacteriophages." *Bacteriophages: Biology, Technology, Therapy* **27**: 19–91.

Sergueev, K. V., A. A. Filippov and M. P. Nikolich (2017). "Highly sensitive bacteriophage-based detection of brucella abortus in mixed culture and spiked blood." *Viruses* **9**(6): 144.

Sharp, N. J., I. J. Molineux, M. A. Page and D. A. Schofield (2016). "Rapid detection of viable bacillus anthracis spores in environmental samples by using engineered reporter phages." *Appl Environ Microbiol* **82**(8): 2380–2387.

Soendergaard, M., J. R. Newton-Northup and S. L. Deutscher (2014). "In vivo phage display selection of an ovarian cancer targeting peptide for SPECT/CT imaging." *Am J Nucl Med Mol Imaging* **4**(6): 561–570.

Stambach, N. R., S. A. Carr, C. R. Cox and K. J. Voorhees (2015). "Rapid detection of listeria by bacteriophage amplification and SERS-lateral flow immunochromatography." *Viruses* **7**(12): 6631–6641.

Steven, A. C., H. L. Greenstone, F. P. Booy, L. W. Black and P. D. Ross (1992). "Conformational changes of a viral capsid protein. Thermodynamic rationale for proteolytic regulation of bacteriophage T4 capsid expansion, co-operativity, and super-stabilization by soc binding." *J Mol Biol* **228**(3): 870–884.

Summers, W. C. (2000). "Felix d'Herelle and the origins of molecular biology." *Journal of the History of Biology* **33**(1):191-194.

Suster, K., A. Podgornik and A. Cor (2017). "Quick bacteriophage-mediated bioluminescence assay for detecting Staphylococcus spp. in sonicate fluid of orthopaedic artificial joints." *New Microbiol* **40**(3): 190–196.

Swift, B. M., T. W. Convery and C. E. Rees (2016). "Evidence of Mycobacterium tuberculosis complex bacteraemia in intradermal skin test positive cattle detected using phage-RPA." *Virulence* **7**(7): 779–788.

Swift, B. M., J. N. Huxley, K. M. Plain, D. J. Begg, K. de Silva, A. C. Purdie, R. J. Whittington and C. E. Rees (2016). "Evaluation of the limitations and methods to improve rapid phage-based detection of viable Mycobacterium avium subsp. paratuberculosis in the blood of experimentally infected cattle." *BMC Vet Res* **12**(1): 115.

Tamariz, J., V. Guevara and H. Guerra (2018). "Rapid detection of salmonellosis due to Salmonella enterica serovar Typhimurium in Peruvian commercially bred cavies, using indigenous wild bacteriophages." *Germs* **8**(4): 178–185.

Thung, T. Y., J. M. Krishanthi Jayarukshi Kumari Premarathne, W. San Chang, Y. Y. Loo, Y. Z. Chin, C. H. Kuan, C. W. Tan, D. F. Basri, C. W. Jasimah Wan Mohamed Radzi and S. Radu (2017). "Use of a lytic bacteriophage to control Salmonella Enteritidis in retail food." *Lwt* **78**: 222–225.

Thung, T. Y., E. Lee, N. A. Mahyudin, K. Anuradha, N. Mazlan, C. H. Kuan, C. F. Pui, F. M. Ghazali, N.-K. Mahmud Ab Rashid, W. D. Rollon, C. W. Tan and S. Radu (2019). "Evaluation of a lytic bacteriophage for bio-control of Salmonella Typhimurium in different food matrices." *Lwt* **105**: 211–214.

Tilton, L., G. Das, X. Yang, N. Wisuthiphaet, I. M. Kennedy and N. Nitin (2019). "Nanophotonic device in combination with bacteriophages for enhancing detection sensitivity of escherichia coli in simulated wash water." *Analytical Letters* **52**(14): 2203–2213.

Turner, D., A. N. Shkoporov, C. Lood, A. D. Millard, B. E. Dutilh, P. Alfenas-Zerbini, L. J. van Zyl, R. K. Aziz, H. M. Oksanen, M. M. Poranen, A. M. Kropinski, J. Barylski, J. R. Brister, N. Chanisvili, R. A. Edwards, F. Enault, A. Gillis, P. Knezevic, M. Krupovic, I. Kurtboke, A. Kushkina, R. Lavigne, S. Lehman, M. Lobocka, C. Moraru, A. Moreno Switt, V. Morozova, J. Nakavuma, A. Reyes Munoz, J. Rumnieks, B. L. Sarkar, M. B. Sullivan, J. Uchiyama, J. Wittmann, T. Yigang and E. M. Adriaenssens (2023). "Abolishment of morphology-based taxa and change to binomial species names: 2022 taxonomy update of the ICTV bacterial viruses subcommittee." *Arch Virol* **168**(2): 74.

Twort, F. W. (1915). "An investigation on the nature of ultra-microscopic viruses." *The Lancet* **186**(4814): 1241–1243.

Vaidya, A., S. Ravindranath and U. S. Annapure (2020). "Detection and differential identification of typhoidal Salmonella using bacteriophages and resazurin." *3 Biotech* **10**(5): 196.

van Helvoort, T. (1992). "Bacteriological and physiological research styles in the early controversy on the nature of the bacteriophage phenomenon." *Med Hist* **36**(3): 243–270.

Wang, C., H. Hang, S. Zhou, Y. D. Niu, H. Du, K. Stanford and T. A. McAllister (2020). "Bacteriophage biocontrol of Shiga toxigenic Escherichia coli (STEC) O145 biofilms on stainless steel reduces the contamination of beef." *Food Microbiol* **92**: 103572.

Wang, D., J. Chen and S. R. Nugen (2017). "Electrochemical detection of escherichia coli from aqueous samples using engineered phages." *Anal Chem* **89**(3): 1650–1657.

Wang, D., T. Hinkley, J. Chen, J. N. Talbert and S. R. Nugen (2019). "Phage based electrochemical detection of Escherichia coli in drinking water using affinity reporter probes." *Analyst* **144**(4): 1345–1352.

Willey, J. M., L. M. Sherwood and C. J. Woolverton (2014). *Prescott's microbiology.* McGraw-Hill.

Xu, J., C. Zhao, Y. Chau and Y. K. Lee (2020). "The synergy of chemical immobilization and electrical orientation of T4 bacteriophage on a micro electrochemical sensor for low-level viable bacteria detection via differential pulse voltammetry." *Biosens Bioelectron* **151**: 111914.

Yan, T., L. Liang, P. Yin, Y. Zhou, A. M. Sharoba, Q. Lu, X. Dong, K. Liu, I. F. Connerton and J. Li (2020). "Application of a novel phage LPSEYT for biological control of salmonella in foods." *Microorganisms* **8**(3): 400.

Yang, X., N. Wisuthiphaet, G. M. Young and N. Nitin (2020). "Rapid detection of Escherichia coli using bacteriophage-induced lysis and image analysis." *PLoS One* **15**(6): e0233853.

Yue, H., Y. He, E. Fan, L. Wang, S. Lu and Z. Fu (2017). "Label-free electrochemiluminescent biosensor for rapid and sensitive detection of pseudomonas aeruginosa using phage as highly specific recognition agent." *Biosens Bioelectron* **94**: 429–432.

Zhang, D., C. P. Coronel-Aguilera, P. L. Romero, L. Perry, U. Minocha, C. Rosenfield, A. G. Gehring, G. C. Paoli, A. K. Bhunia and B. Applegate (2016). "The use of a novel NanoLuc -based reporter phage for the detection of escherichia coli O157:H7." *Sci Rep* **6**: 33235.

Zhang, H., R. Wang and H. Bao (2013). "Phage inactivation of foodborne Shigella on ready-to-eat spiced chicken." *Poult Sci* **92**(1): 211–217.

Zhou, Y., A. Marar, P. Kner and R. P. Ramasamy (2017). "Charge-directed immobilization of bacteriophage on nanostructured electrode for whole-cell electrochemical biosensors." *Anal Chem* **89**(11): 5734–5741.

17 Plant Viruses

Rashmi Singh, Sachin Kumar and Latha Rangan
Department of Biosciences and Bioengineering, Indian
Institute of Technology Guwahati, Assam, India

ABBREVIATIONS

BMV: brome mosaic virus, CaMV: cauliflower mosaic virus, dsDNA: double-stranded DNA, ICTV: International Committee on Taxonomy of Viruses, PVNPs: plant virus nanoparticles, PVX: potato virus X, RNAi: RNA interference, ssRNA: single-stranded RNA, ssDNA: single-stranded DNA, TMV: tobacco mosaic virus, VIGS: virus-induced gene silencing, VLPs: viral-like particles.

INTRODUCTION

In 1898, the term "virus" was first attributed to a plant pathogen known as tobacco mosaic virus (TMV), which infects a broad spectrum of plants, leading to a characteristic yellow mosaic pattern on their leaves. Since this initial discovery, research into plant viruses has contributed to significant advancements in both virology and plant biology, resulting in the emergence of novel findings (Xavier and Whitfield, 2023). In 1935, Wendell Stanley, a scientist from the USA, isolated TMV in a crystalline form by processing 4,000 kg of infected tobacco leaves with ammonium sulfate. He found the material to be highly infectious and identified it as a protein. The following year, UK scientists Frederick Bawden and Norman Pirie purified TMV and discovered it contained carbohydrates and phosphorus, characterizing it as a liquid crystalline nucleoprotein. In 1939, German scientists produced the first images of TMV particles using an electron microscope. This breakthrough paved the way for the first-ever image of a virion to be captured in 1939, revealing TMV. Then, in 1957, Fraenkel-Conrat and colleagues showed that RNA can act as genetic material like DNA, employing TMV in a model system, tobacco plants. For their contributions, Wendell M. Stanley and his colleagues were later awarded the Nobel Prize in Chemistry in 1946 (Wang et al., 2020; Wilson, 2014). Among the myriad disease-causing agents affecting plants, viruses pose a significant threat to agricultural productivity. They have been identified as the cause of approximately half of the newly emerging infectious diseases observed in various crop plants, resulting in approximately 40% of overall crop losses due to infection. Across the globe, over 25 plant virus families have been documented to infect a diverse array of crop species, leading to substantial economic losses (Mehetre et al., 2021). In the early 1990s, studies hinted at a defense system in plants that target viral RNA very precisely. This system, now known as RNA interference (RNAi), was initially not well understood but was recognized as distinct from other defense systems involving proteins. Over time, scientists have studied the intricate interactions between viruses and the host RNAi system. Additionally, it was revealed that plant viruses play a multifaceted role, not just causing diseases but also serving as tools to study important cellular processes such as how materials move within and between cells, as well as how genes are regulated (Nelson and Citovsky, 2005). This chapter explores plant viruses and their recent progress in research, which has been instrumental in biotechnology applications across medicine, the materials industry, and agriculture.

DOI: 10.1201/9781003369349-17

UNDERSTANDING PLANT VIRUS DIVERSITY

Viruses have much more genetic variety than the cells that make up living things. Their widespread presence throughout history suggests that instead of just causing diseases, viruses have probably played important roles in ecosystems worldwide since life began (Lefeuvre et al., 2019). Viruses are parasites that infect and replicate within cellular organisms. Among them are viral agents that target vertebrates, invertebrates, plants, fungi, and bacteria. However, there are exceptions, such as *Tospoviruses* and certain members of the *Rhabdoviridae* family, which can infect both plants and invertebrates, blurring the line between plant and animal viruses (Hull, 2013). Plant viruses are responsible for many damaging plant diseases globally, resulting in significant losses in crop quality and yield. Infected plants typically display symptoms such as yellowing leaves, striped or blotchy patterns on leaves, curled leaves, stunted growth, and abnormalities in flower or fruit formation (Awasthi et al., 2016).

Plant viruses come in various structures and ecological roles. They are grouped into two categories based on their genetic makeup: *Monodnaviria*, containing ssDNA viruses, and *Riboviria*, housing RNA viruses. RNA viruses can have single or double strands, and within *Riboviria*, there are caulimoviruses, dsDNA viruses using reverse transcription. RNA viruses are further categorized as positive or negative sense, based on their genetic orientation. Most plant viruses fall under the positive-sense ssRNA category (Xavier and Whitfield, 2023).

CLASSIFICATION OF PLANT VIRUSES

In 1966, the International Committee on Nomenclature of Viruses (ICNV) was formed, comprising 43 virologists worldwide. Their goal was to establish a unified classification system for viruses. By 1970, the ICNV issued its first classification, including plant viruses, emphasizing virus particle properties as the primary grouping criteria, marking a pivotal moment in virus taxonomy (Pagán, 2018).

Currently, over 60 characteristics are employed for classifying viruses (Awasthi et al., 2016). Originally, virus classification and taxon nomenclature relied on observable properties such as virion morphology, nucleic acid type, and physical characteristics like susceptibility to environmental factors. While genomic characterization is now a prerequisite, debate persists on the ongoing relevance of traditional criteria and whether *in vitro* culture or virion visualization is necessary for taxonomic assignment (Simmonds et al., 2023). These characteristics aid in placing viruses into genera and higher taxa. Additionally, distinguishing between related viruses and assessing degrees of relationship within a group requires consideration of properties with many variants, such as symptoms, host range, nucleotide sequence, and amino acid composition of specific proteins like the coat protein. Serological specificity and amino acid sequences can also serve to define groups and differentiate viruses within them (Hull, 2013). In April 2023, the International Committee on Taxonomy of Viruses (ICTV) approved and certified changes to virus taxonomy, resulting in updates to 72 orders, 8 suborders, 264 families, 182 subfamilies, 2,818 genera, 84 subgenera, and 11,273 species (Zerbini et al., 2023). Currently, ICTV has documented 131 families comprising 803 genera and 4,853 species of plant viruses (Rahman and Sanan-Mishra, 2024). According to preliminary research on wild plants, many more plant viruses have yet to be identified (Chandra and Awasthi, 2020).

STRUCTURE OF PLANT VIRUSES: COMPOSITION AND MORPHOLOGY

Plant viruses typically possess small genomes, ranging from 4 to 20 kilobases (kb), which encode approximately 5 to 10 proteins. Recent research has shed light on the complexity of these compact genomes and the multifunctional nature of plant virus proteins (Sanfaçon, 2017). The structural components of virus particles include the capsid, which is the protein shell enclosing the virus, and capsomeres, clusters of coat-protein subunits visible under electron microscopy. Certain viruses

Plant Viruses

possess an envelope, a lipoprotein membrane surrounding the nucleocapsid, the inner nucleoprotein core of the virus. The nucleocapsid consists of genomic nucleic acid bound to viral proteins but not yet formed into a mature virion. The mature virus particle, known as a virion or virus particle, includes nucleic acid and capsid, and in some cases, a lipoprotein membrane. These components are vital for virus replication and infection (Wilson, 2014). Viral capsids, which encase viral genomes, can be icosahedral or helical in structure. Helical capsids are found in various viruses, including filamentous bacteriophages, nucleocapsids of negative-stranded RNA genomes, and many plant viruses. Notably, helical capsids are present in about half of plant viruses known to have positive-stranded RNA genomes, despite icosahedral capsids being considered more efficient in RNA packaging. Helical capsids allow for longer RNA genomes, providing functional advantages for viruses with larger genomes. In general, these capsids consist of many identical viral capsid protein subunits, with a single viral genomic RNA molecule enclosed in each virus particle. Rod-shaped and flexuous filamentous viruses are two more subtypes of helical plant viruses, the latter being more prevalent (Solovyev and Makarov, 2016). Presently, cryo-electron microscopy stands as the forefront microscopy method for studying plant–virus interactions, offering the capability to show structures at near-atomic levels (Mehetre et al., 2021).

MODES OF TRANSMISSION OF PLANT VIRUSES

Plant viruses exhibit four distinct lifestyles: persistent, acute, chronic, and endogenous. While some viruses can transition between lifestyles, particularly between acute and chronic, such changes are rare. Persistent viruses typically show no symptoms and are transmitted vertically via gametes, remaining within cells, including the meristem, without horizontal movement. Acute viruses, however, are often transmitted horizontally, move readily between cells, and accumulate to high levels in plants, flourishing in monoculture. Infections by acute viruses can result in host death, recovery, or conversion to chronic infection. Chronic viruses persist within host plants for extended periods, potentially without causing visible symptoms. Additionally, endogenous viruses are integrated into the plant genome, mostly as defective remnants of ancient infections, but some can be activated under specific conditions (Roossinck, 2010).

HOST–VIRUS INTERACTIONS

The interactions between viruses and plants vary in the mechanisms that lead to symptom development, with both compatible and incompatible host–virus interactions occurring during viral infections in plants (Yadav and Chhibbar, 2018).

COMPATIBLE HOST–VIRUS RELATIONSHIPS

When a virus infects a host cell, it can lead to symptoms appearing locally or systemically, or both simultaneously, known as external symptoms. These symptoms, such as leaf rolling, withering, yellowing, stunting, necrosis, and the creation of mosaic patterns, characterize viral infections in plants.

INCOMPATIBLE HOST–VIRUS RELATIONSHIPS

Once a host plant recognizes a virus, it sets off a series of defense mechanisms that prevent the virus from replicating and moving within the host, leading to incompatible host–virus associations. Extreme immunity in host plants prevents virus infection, leading to the absence of external symptoms and undetectable viruses. These interactions may arise due to various factors, including the lack of crucial cellular components necessary for virus replication or movement, the presence of antiviral defense mechanisms, or a combination of these factors (Garcia-Ruiz, 2019). Host plant resistance to viruses can be qualitative, involving a specific relationship between viral and host resistance genes resulting in hypersensitivity or resistance to virus spread, or quantitative, where no specific gene relationship is observed, manifesting as resistance to the spread and multiplication of viruses, field resistance, or tolerance to plant disease (Marwal and Gaur, 2020). The connection between plant viruses and their hosts is typically seen as parasitic, often causing harm and disease to the host plants. This viewpoint has influenced how plant viruses are classified, focusing on those that visibly affect cultivated crops. However, recent research has shown that some plants can carry viruses without showing any signs of illness. Advanced methods like metagenomics have unveiled that these asymptomatic infections are more widespread. They can happen because plants either tolerate the virus or the virus persists at low levels to prevent damage to the host (Hasiów-Jaroszewska et al., 2021; Mauck et al., 2012). The interaction between viruses and their plant hosts may cause the host to develop an antiviral response. Despite the seemingly simple genome organization of most plant viruses, this interaction is a dynamic and intricate process. It involves several interactions among viral-coded and host-coded proteins, as well as interactions between proteins and viral nucleic acids, and proteins and host membranes (lipids). Studying these interactions over time is essential for gaining a deeper understanding of plant virus infections and their impact on host cells (Nagy, 2008; Norberg et al., 2023).

SIGNIFICANT PLANT VIRUSES FOR RESEARCH AND STUDY

Some of the most scientifically significant viruses include TMV, cucumber mosaic virus (CMV), potato virus X (PVX), brome mosaic virus (BMV), and cauliflower mosaic virus (CaMV). Among these, TMV and CMV are particularly notable, as indicated by their prominence in the research field (Scholthof et al., 2011).

Tobacco Mosaic Virus (TMV)

Tobacco mosaic virus holds a significant place in plant virology and is recognized as one of the most important plant viruses. Initially identified by Martinus Beijerinck in 1898, TMV was later confirmed as a virus by Henry A. Allard in 1916. Subsequent research by Helen Purdy Beale, Francis O. Holmes, and Howard H. McKinney introduced essential tools for virology, including serology and cross-protection techniques (Wang et al., 2020).

The importance of TMV extends beyond its economic impact on tobacco crops; it has been pivotal in scientific breakthroughs, leading to the Nobel Prize and pioneering discoveries such as the first plant virus RNA sequencing and the demonstration of gene-for-gene resistance. Despite early misconceptions about its structure, the function of TMV in replication and movement has been elucidated through extensive research. Its utility extends to various fields, including biotechnology, computer data storage, and understanding host–virus interactions at the cellular level (Scholthof, 2024).

Cucumber Mosaic Virus (CMV)

Cucumber mosaic virus belongs to the *Cucumovirus* genus within the *Bromoviridae* family, exhibits an extensive host range, and is widespread globally. It has been identified as a causative agent for various diseases affecting vegetable and pulse crops, ornamental plants, medicinal herbs, and weeds (Joshi et al., 2023). It has 29 nm diameter icosahedral particles, composed of 180 subunits of a single coat protein (CP) and genomic RNAs. CMV strains are categorized into two major subgroups, I and II, with further divisions within subgroup I. The CMV genome encompasses five genes expressed from genomic and subgenomic RNAs. Proteins 1a and 2a participate in virus replication, while protein 2b suppresses RNA silencing and facilitates viral recombination. All CMV-encoded proteins regulate the long-distance and cell-to-cell migration of protein 3a and CP. CMV

Plant Viruses

interacts synergistically with various viruses and has a broad host range, infecting over 1,200 plant species. Despite control challenges, pathogen-derived resistance holds promise for managing CMV (Palukaitis and García-Arenal, 2018).

APPLICATIONS OF PLANT VIRUSES IN BIOTECHNOLOGY

In recent years, plant viruses have received attention in biotechnology for diverse applications (Table 17.1). Their ability to trigger the innate immune system via pathogen-associated molecular pattern (PAMP) receptors makes them ideal for vaccine production, especially as they pose no harm

TABLE 17.1
Plant Viruses and Their Application in Biotechnology

Virus	Nucelic Acid Type	Family	Applications in Biotechnology	Reference
Tobacco mosaic virus (TMV)	ssRNA	*Tombusviridae*	Molecular imaging and theragnostic, virus-induced resistance, tissue engineering, biosensing and diagnostics, cancer treatment	(Venkataraman and Hefferon, 2021)
Potato virus X (PVX)	ssRNA	*Potexviridae*	Molecular imaging, immunotherapy, biosensing and diagnostics, biocatalyst generation, cancer chemotherapy	(Le et al., 2017; Venkataraman and Hefferon, 2021)
Cowpea mosaic virus (CPMV)	ssRNA	*Comoviridae*	Immunomodulation, cancer therapy, nanoparticle applications, *in situ* vaccine (ISV) for tumors, vaccine production	(Beiss et al., 2022; Venkataraman and Hefferon, 2021)
Bean yellow dwarf virus (BeYDV)	ssDNA	*Geminiviridae*	Expression of pharmaceutical proteins, plant-made recombinant immune complex (RIC) vaccines, humanized glycan profile and genome editing, virus-induced gene silencing (VIGS)	(Yang QiuYing et al., 2017)
Brome mosaic virus (BMV)	ssRNA	*Bromoviridae*	VIGS, therapeutic delivery, nanobioreactors, cancer treatment, pharmaceutical development, immunogenicity studies, magnetic imaging and biosensing	(He et al., 2021; Lomonossoff and Evans, 2011; Villanueva-Flores et al., 2023; Young et al., 2008)
Carnation mottle virus (CarMV)	ssRNA	*Tombusviridae*	Nanomaterials, 2D/3D patterning/array formation	(Young et al., 2008)

(*Continued*)

TABLE 17.1 (CONTINUED)
Plant Viruses and Their Application in Biotechnology

Virus	Nucelic Acid Type	Family	Applications in Biotechnology	Reference
Cowpea chlorotic mottle virus (CCMV)	ssRNA	*Bromoviradae*	Nanomaterials, polymer loading/encapsidation, chemical conjugation, biomedical applications, imaging agents, targeting biodistribution, delivery of heterologous nucleic acids, gene therapy, *in situ* cancer vaccine development	(Phelps et al., 2007; Tscheuschner et al., 2023; Young et al., 2008)
Red clover necrotic mosaic virus (RCNMV)	ssRNA	*Tombusviridae*	Nanomaterials, Au, CoFe2O4, and CdSe nanoparticles, biomedical applications, doxorubicin/fluorophore infusion, targeted particles for cancer treatment	(Lomonossoff and Evans, 2011; Young et al., 2008)
Turnip yellow mosaic virus (TYMV)	ssRNA	*Tymoviridae*	Biomedical applications, fluorescent labeling/sensor development, nanonets immobilizing industrial enzyme formation, tissue engineering	3, 10, 13 (Fermin et al., 2018; Nellist et al., 2022; Young et al., 2008)
Cucumber mosaic virus (CMV)	ssRNA	*Bromoviridae*	Genome editing, nanotechnology, delivery of therapeutic agents, used in cat allergy, psoriasis, insect bite, and atopic dermatitis	(Kim et al., 2020; Lomonossoff and Evans, 2011; Zeltins et al., 2017)
Barley stripe mosaic virus (BSMV)	ssRNA	*Virgaviridae*	Biotemplate for nanomaterial synthesis, nanoparticle synthesis, VIGS, vascular puncture inoculation (VPI)	(Clare et al., 2015; Kant and Dasgupta, 2019; Lee et al., 2020)
Cauliflower mosaic virus (CaMV)	dsDNA	*Caulimoviridae*	Targeted drug delivery, epitope carriers for vaccines, gene delivery carriers, immunostimulation,	(Agrios, 2005; Pouresmaeil et al., 2023)

to mammals. Furthermore, they can induce both cell-mediated and humoral immune responses when administered through mucosal or parenteral routes (Venkataraman and Hefferon, 2021). Plant viruses serve as vectors, facilitating transient expression of heterologous proteins like vaccine antigens and antibodies. Plant viral vectors significantly cut down on costs and time associated with regulating gene expression in comparison to conventional transgenic methods, making them highly promising for agricultural and medical applications. Moreover, vectors have been engineered to generate both native and foreign proteins, enhancing agricultural characteristics and facilitating vaccine production (Wang et al., 2020). Moreover, they enable the production of therapeutic

Plant Viruses

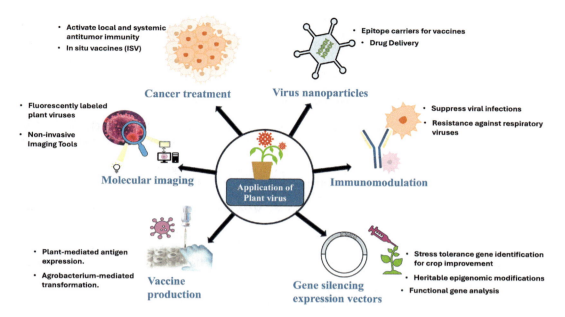

FIGURE 17.1 Application of plant viruses in medicine and technology.

proteins for human or animal use and allow for targeted exploration of plant biochemical processes. In medicine, plant virus nanoparticles (PVNPs) are used for vaccine production, targeted drug delivery, as epitope carriers in cancer immunotherapy, gene delivery, and medical imaging (Figure 17.1) (Pouresmaeil et al., 2023).

Plant viruses exhibit several advantageous properties for bionanotechnology applications. They are not considered a biological concern and are safe for other organisms. High expression yields are obtained from the efficient and quick production of viral particles, particularly when natural host systems are used. Additionally, heterologous expression systems can generate abundant virus-like particles (VLPs). Moreover, viral capsids are remarkably resilient, withstanding high temperatures, a wide pH range, and even organic solvent–water mixtures, making them versatile building blocks for diverse applications in bionanotechnology (Steinmetz and Evans, 2007).

CONCLUSION

Plant virology continues to unravel the complexities of virus–host interactions, offering insights into plant diseases and biotechnological innovations. With technological advancements and research methodologies, the field holds promise for addressing global challenges in agriculture and healthcare. In conclusion, the growing awareness of plant viruses in biotechnology and medicine highlights their significant potential in various applications. Their ability to trigger immune responses without posing harm to mammals makes them promising candidates for vaccine production. Additionally, they serve as vectors that make it easier to produce therapeutic medicines and promote the expression of heterologous proteins. The use of PVNPs in targeted drug delivery, cancer immunotherapy, gene delivery, and medical imaging highlights their versatility in medicine. Overall, the broad utility of plant viruses in diverse fields underscores their importance as innovative tools in biotechnology and medicine.

REFERENCES

Agrios, G.N., 2005. Plant diseases caused by fungi. *Plant Pathology* 385–614.

Awasthi, S., Chauhan, R., Narayan, R.P., 2016. Plant viruses: History and taxonomy. In: Gaur, R.K., Petrov, N.M., Patil, B.L., Stoyanova, M.I. (Eds.), *Plant Viruses: Evolution and Management*. Springer Singapore, pp. 1–17. https://doi.org/10.1007/978-981-10-1406-2_1

Beiss, V., Mao, C., Fiering, S.N., Steinmetz, N.F., 2022. Cowpea mosaic virus outperforms other members of the secoviridae as in situ vaccine for cancer immunotherapy. *Molecular Pharmaceutics* 19, 1573–1585. https://doi.org/10.1021/acs.molpharmaceut.2c00058

Chandra, P., Awasthi, L.P., 2020. Plant virus taxonomy. In: Awasthi, L.P., *Applied plant virology*. Elsevier, pp. 421–434.

Clare, D.K., Pechnikova, E.V., Skurat, E.V., Makarov, V.V., Sokolova, O.S., Solovyev, A.G., Orlova, E.V., 2015. Novel inter-subunit contacts in barley stripe mosaic virus revealed by cryo-electron microscopy. *Structure* 23, 1815–1826.

Fermin, G., Rampersad, S., Tennant, P., 2018. Viruses as tools of biotechnology: therapeutic agents, carriers of therapeutic agents and genes, nanomaterials, and more. *Viruses: Molecular Biology, Host Interactions, and Applications to Biotechnology,* 291–316.

Garcia-Ruiz, H., 2019. Host factors against plant viruses. *Molecular Plant Pathology* 20, 1588–1601. https://doi.org/10.1111/mpp.12851

Hasiów-Jaroszewska, B., Boezen, D., Zwart, M.P., 2021. Metagenomic studies of viruses in weeds and wild plants: a powerful approach to characterise variable virus communities. *Viruses* 13, 1939.

He, G., Zhang, Z., Sathanantham, P., Diaz, A., Wang, X., 2021. Brome mosaic virus (Bromoviridae). *Encyclopedia of Virology* 252–259.

Hull, R., 2013. *Plant virology*. Academic Press.

Joshi, M., Narute, T.K., Sarnobat, D., 2023. The cucumber mosaic virus: A review. *International Journal of Plant & Soil Science* 35, 253–260.

Kant, R., Dasgupta, I., 2019. Gene silencing approaches through virus-based vectors: Speeding up functional genomics in monocots. *Plant Molecular Biology* 100, 3–18. https://doi.org/10.1007/s11103-019-00854-6

Kim, H., Onodera, Y., Masuta, C., 2020. Application of cucumber mosaic virus to efficient induction and long-term maintenance of virus-induced gene silencing in spinach. *Plant Biotechnology* 37, 83–88.

Le, D.H., Lee, K.L., Shukla, S., Commandeur, U., Steinmetz, N.F., 2017. Potato virus X, a filamentous plant viral nanoparticle for doxorubicin delivery in cancer therapy. *Nanoscale* 9, 2348–2357.

Lee, K.Z., Pussepitiya, V.B., Lee, Y.H., Fries, S.L., Harris, M., Hemmati, S., Solomon, K., 2020. Engineering tobacco mosaic virus, barley stripe mosaic virus, and their virus-like-particles for synthesis of biotemplated nanomaterials. https://doi.org/10.22541/au.159414778.89024447

Lefeuvre, P., Martin, D.P., Elena, S.F., Shepherd, D.N., Roumagnac, P., Varsani, A., 2019. Evolution and ecology of plant viruses. *Nature Reviews Microbiology* 17, 632–644.

Lomonossoff, G.P., Evans, D.J., 2011. Applications of plant viruses in bionanotechnology. In: Palmer, K., Gleba, Y. (Eds.), *Plant Viral Vectors, Current Topics in Microbiology and Immunology*. Springer, pp. 61–87. https://doi.org/10.1007/82_2011_184

Marwal, A., Gaur, R.K., 2020. Host plant strategies to combat against viruses effector proteins. *Current Genomics* 21, 401–410.

Mauck, K., Bosque-Pérez, N.A., Eigenbrode, S.D., De Moraes, C.M., Mescher, M.C., 2012. Transmission mechanisms shape pathogen effects on host–vector interactions: evidence from plant viruses. *Functional Ecology* 26, 1162–1175. https://doi.org/10.1111/j.1365-2435.2012.02026.x

Mehetre, G.T., Leo, V.V., Singh, G., Sorokan, A., Maksimov, I., Yadav, M.K., Upadhyaya, K., Hashem, A., Alsaleh, A.N., Dawoud, T.M., 2021. Current developments and challenges in plant viral diagnostics: A systematic review. *Viruses* 13, 412.

Nagy, P.D., 2008. Yeast as a Model Host to Explore Plant Virus-Host Interactions. *Annual Review of Phytopathology* 46, 217–242. https://doi.org/10.1146/annurev.phyto.121407.093958

Nellist, C.F., Ohshima, K., Ponz, F., Walsh, J.A., 2022. Turnip mosaic virus, a virus for all seasons. *Annals of Applied Biology* 180, 312–327. https://doi.org/10.1111/aab.12755

Nelson, R.S., Citovsky, V., 2005. Plant viruses. Invaders of cells and pirates of cellular pathways. *Plant Physiology* 138, 1809–1814.

Norberg, A., Susi, H., Sallinen, S., Baran, P., Clark, N.J., Laine, A.-L., 2023. Direct and indirect viral associations predict coexistence in wild plant virus communities. *Current Biology* 33, 1665–1676.

Pagán, I., 2018. The diversity, evolution and epidemiology of plant viruses: A phylogenetic view. *Infection, Genetics and Evolution* 65, 187–199.

Palukaitis, P., García-Arenal, F., 2018. *Cucumber mosaic virus*. American Phytopathological Society (APS Press).

Phelps, J.P., Dao, P., Jin, H., Rasochova, L., 2007. Expression and self-assembly of cowpea chlorotic mottle virus-like particles in Pseudomonas fluorescens. *Journal of Biotechnology* 128, 290–296.

Pouresmaeil, M., Dall'Ara, M., Salvato, M., Turri, V., Ratti, C., 2023. Cauliflower mosaic virus: Virus-host interactions and its uses in biotechnology and medicine. *Virology* 580, 112–119.

Rahman, A., Sanan-Mishra, N., 2024. When an intruder comes home: GM and GE strategies to combat virus infection in plants. *Agriculture* 14, 282.

Roossinck, M.J., 2010. Lifestyles of plant viruses. *Philosophical Transactions of the Royal Society B* 365, 1899–1905. https://doi.org/10.1098/rstb.2010.0057

Sanfaçon, H., 2017. Grand challenge in plant virology: Understanding the impact of plant viruses in model plants, in agricultural crops, and in complex ecosystems. *Frontiers in Microbiology* 8, 268638.

Scholthof, K.-B.G., 2024. *Tobacco mosaic virus: The beginning of plant virology.* Phytopathology News.

Scholthof, K.-B.G., Adkins, S., Czosnek, H., Palukaitis, P., Jacquot, E., Hohn, T., Hohn, B., Saunders, K., Candresse, T., Ahlquist, P., Hemenway, C., Foster, G.D., 2011. Top 10 plant viruses in molecular plant pathology. *Molecular Plant Pathology* 12, 938–954. https://doi.org/10.1111/j.1364-3703.2011.00752.x

Simmonds, P., Adriaenssens, E.M., Zerbini, F.M., Abrescia, N.G., Aiewsakun, P., Alfenas-Zerbini, P., Bao, Y., Barylski, J., Drosten, C., Duffy, S., 2023. Four principles to establish a universal virus taxonomy. *PLoS Biology* 21, e3001922.

Solovyev, A.G., Makarov, V.V., 2016. Helical capsids of plant viruses: architecture with structural lability. *Journal of General Virology* 97, 1739–1754. https://doi.org/10.1099/jgv.0.000524

Steinmetz, N.F., Evans, D.J., 2007. Utilisation of plant viruses in bionanotechnology. *Organic & Biomolecular Chemistry* 5, 2891–2902.

Tscheuschner, G., Ponader, M., Raab, C., Weider, P.S., Hartfiel, R., Kaufmann, J.O., Völzke, J.L., Bosc-Bierne, G., Prinz, C., Schwaar, T., 2023. Efficient purification of cowpea chlorotic mottle virus by a novel peptide aptamer. *Viruses* 15, 697.

Venkataraman, S., Hefferon, K., 2021. Application of plant viruses in biotechnology, medicine, and human health. *Viruses* 13, 1697.

Villanueva-Flores, F., Pastor, A.R., Palomares, L.A., Huerta-Saquero, A., 2023. A novel formulation of asparaginase encapsulated into virus-like particles of brome mosaic virus: In vitro and in vivo evidence. *Pharmaceutics* 15, 2260.

Wang, M., Gao, S., Zeng, W., Yang, Y., Ma, J., Wang, Y., 2020. Plant virology delivers diverse toolsets for biotechnology. *Viruses* 12, 1338.

Wilson, C.R., 2014. *Applied plant virology.* CABi.

Xavier, C.A., Whitfield, A.E., 2023. Plant virology. *Current Biology* 33, R478–R484.

Yadav, S., Chhibbar, A.K., 2018. Plant–virus interactions. In: Singh, A., Singh, I.K. (Eds.), *Molecular aspects of plant-pathogen interaction.* Springer Singapore, pp. 43–77. https://doi.org/10.1007/978-981-10-7371 -7_3

Yang QiuYing, Y.Q., Ding Bo, D.B., Zhou XuePing, Z.X., 2017. Geminiviruses and their application in biotechnology. *Journal of Integrative Agriculture*, 16(12), 2761–2771. https://doi.org/10.1016/S2095 -3119(17)61702-7

Young, M., Debbie, W., Uchida, M., Douglas, T., 2008. Plant viruses as biotemplates for materials and their use in nanotechnology. *Annual Review of Phytopathology* 46, 361–384. https://doi.org/10.1146/annurev .phyto.032508.131939

Zeltins, A., West, J., Zabel, F., El Turabi, A., Balke, I., Haas, S., Maudrich, M., Storni, F., Engeroff, P., Jennings, G.T., 2017. Incorporation of tetanus-epitope into virus-like particles achieves vaccine responses even in older recipients in models of psoriasis, Alzheimer's and cat allergy. *NPJ Vaccines* 2, 30.

Zerbini, F.M., Siddell, S.G., Lefkowitz, E.J., Mushegian, A.R., Adriaenssens, E.M., Alfenas-Zerbini, P., Dempsey, D.M., Dutilh, B.E., García, M.L., Hendrickson, R.C., Junglen, S., Krupovic, M., Kuhn, J.H., Lambert, A.J., Łobocka, M., Oksanen, H.M., Robertson, D.L., Rubino, L., Sabanadzovic, S., Simmonds, P., Smith, D.B., Suzuki, N., Van Doorslaer, K., Vandamme, A., Varsani, A., 2023. Changes to virus taxonomy and the ICTV Statutes ratified by the international committee on taxonomy of viruses (2023). *Archives of Virology* 168, 175. https://doi.org/10.1007/s00705-023-05797-4

18 Oncolytic Viruses

Deepa Mehta and Sachin Kumar
Department of Biosciences and Bioengineering, Indian
Institute of Technology Guwahati, Guwahati, Assam, India

INTRODUCTION

The term "oncolytic virus" is derived from the Greek words *onco* (tumor) and *lysis* (breakdown or destruction). Oncolytic viruses (OVs) are a promising therapeutic strategy for the treatment of cancer since they target and kill cancer cells only, leaving healthy cells unharmed. In the early 1900s, physician John Beard proposed that viruses might trigger an immune response against cancer. It was the first time that the concept of utilizing viruses to cure cancer was raised. Chester Southam, a researcher, administered live virus injections to cancer patients in the 1950s, causing the tumors to recede without affecting healthy tissue. However, the lack of necessary control and regulation made the study contentious, resulting in a backlash from society.

Interest in oncolytic viruses was rekindled in the 1990s, as advances in genetic engineering and molecular biology made it possible to tweak viruses to lessen their virulence, boost their selectivity for cancer cells, and improve their anti-tumor effects. A study from the National Cancer Institute (NCI) (Martuza et al., 1991) showed that a genetically modified herpes simplex virus was able to kill tumor cells in mice without significantly harming normal tissue. The basic principle behind oncolytic virus therapy is that the virus can replicate specifically within tumor cells, causing them to burst and release viral particles that can then infect and destroy neighboring cancer cells. This process is known as "oncolysis." In addition to direct tumor cell killing, oncolytic viruses can also stimulate the immune system to attack cancer cells, enhancing the overall anti-cancer effect of the therapy.

Since then, many oncolytic viruses have been created and examined in preclinical and clinical investigations, with some exhibiting encouraging outcomes in early-phase trials. Talimogene laherparepvec (T-VEC) was the first oncolytic virus authorized by the US Food and Drug Administration (FDA) in 2015 for the treatment of melanoma (Andtbacka et al., 2015). Herpes simplex virus type 1 (HSV-1) is one of the most studied oncolytic viruses and has been utilized for the treatment of melanoma, glioma, and other malignancies (Hua et al., 2019). Adenoviruses have undergone significant research and have shown efficacy against prostate cancer and other solid tumors (Patel et al., 2009). Table 18.1 includes a list of additional viruses researched for their potential as oncolytic agents. The use of combination therapies involving oncolytic viruses and other cancer treatments like chemotherapy and immunotherapy is currently being researched. Although there have been positive outcomes in clinical trials, more investigation is required to improve the delivery and effectiveness of oncolytic viruses, understand their mechanisms of action, and determine the optimal approach for combining them with other therapies.

BIOLOGY AND MECHANISM OF ONCOLYTIC VIRUSES

VIRAL SELECTION AND MODIFICATION

The biology of oncolytic viruses involves their interaction with tumor cells, replication within the tumor microenvironment, and the mechanisms through which they induce tumor cell death.

TABLE 18.1
Common Oncolytic Viruses

Oncolytic Virus	Genome	Targeted Cancer	Clinical Trial Phase
Imlygic (talimogene laherparepvec/T-VEC)		Melanoma lesions in the skin and lymph nodes	FDA approved
RIGVIR (Riga virus)		Melanoma	FDA approved
Reovirus	ds RNA virus	Head and neck cancer	FDA approved
Adenovirus	dsDNA	Various solid tumors	Phase III
Herpes simplex virus	dsDNA	Melanoma, glioblastoma	Phase III
Adenovirus serotype 5	dsDNA	Prostate cancer	Preclinical
Herpes simplex virus type 1	dsDNA	Bladder cancer	Preclinical
Vaccinia virus	dsDNA	Various solid tumors	Phase II
Adeno-associated virus	ssDNA	Hepatocellular carcinoma	Phase I/II
Adeno-associated virus type 2	ssDNA	Glioblastoma	Preclinical
Parvovirus	ssDNA	Glioblastoma	Phase I/II
Newcastle disease virus	ssRNA	Breast cancer, colon cancer, lung cancer, melanoma	Phase II
Vesicular stomatitis virus	ssRNA	Liver cancer, pancreatic cancer, colon cancer, lung cancer, melanoma	Phase II
Measles virus	ssRNA	Multiple myeloma, ovarian cancer, glioblastoma	Phase I/II
Coxsackievirus	ssRNA	Pancreatic cancer	Phase I/II
Seneca Valley virus	ssRNA	Small cell lung cancer	Phase I/II
Maraba virus	ssRNA	Various solid tumors	Phase I/II
Poliovirus	ssRNA	Glioblastoma	Phase I/II
Mumps virus	ssRNA	Pancreatic cancer	Phase I/II
Oncolytic retrovirus	ssRNA	Various solid tumors	Phase I/II
Lentiviral vector	ssRNA	Glioblastoma	Phase I/II
Sendai virus	ssRNA	Breast cancer	Phase I/II
Zika virus	ssRNA	Glioblastoma	Preclinical
Vesicular stomatitis virus (VSV)	ssRNA	Glioblastoma	Preclinical
CoxsackievirusB3	ssRNA	Glioblastoma	Preclinical
Echovirus 1	ssRNA	Glioblastoma	Preclinical
Sindbis virus	ssRNA	Breast cancer	Preclinical
Western equine encephalitis virus	ssRNA	Glioblastoma	Preclinical
Parainfluenza virus type 2	ssRNA	Lung cancer	Preclinicall
Semliki Forest virus	ssRNA	Breast cancer	Preclinical

- Selective tumor cell infection: Oncolytic viruses identify unique markers such as overexpressed receptors or dysregulated signaling pathways to target and enter tumor cells while sparing healthy cells.
- Viral replication: Once inside tumor cells, viruses utilize cellular machinery to replicate their genetic material and produce viral progeny. These viral progeny are released upon tumor cell lysis, facilitating the spread of the virus to neighboring cancer cells within the tumor microenvironment.
- Tumor cell killing mechanisms: Oncolytic viruses employ various mechanisms to induce tumor cell death, which are discussed in the following section.
- Immune response activation: Oncolytic viruses can stimulate direct tumor cell killing, initiate innate immune responses, and activate adaptive immune responses.

- Anti-tumor immunity and systemic effects: The immune response triggered by oncolytic viruses can result in the generation of a systemic anti-tumor immune response. Immune cells, such as cytotoxic T cells, can recognize tumor-specific antigens and target cancer cells both at the site of viral infection and at distant metastatic sites, leading to a broader anti-tumor effect.

TUMOR CELL KILLING MECHANISMS

Oncolytic viruses can kill cancer cells through a variety of mechanisms. Some of the main mechanisms of action include:

- Direct tumor cell lysis: Oncolytic viruses can infect and replicate within cancer cells, causing them to rupture and die.
- Induction of apoptosis: Some oncolytic viruses can induce cancer cells to undergo programmed cell death, or apoptosis.
- Activation of immune response: Oncolytic viruses can initiate innate immune responses, including the production of pro-inflammatory cytokines and chemokines, which recruit immune cells to the tumor microenvironment. Additionally, they can enhance the presentation of tumor antigens to immune cells, facilitating the activation of adaptive immune responses.
- Anti-angiogenic effects: Oncolytic viruses can inhibit the growth of blood vessels that supply nutrients to tumors, thereby depriving the tumor of oxygen and nutrients necessary for its survival (Hou et al., 2014).
- Combinatorial effects: Oncolytic viruses can be used in combination with other cancer treatments such as chemotherapy, radiation therapy, or immunotherapy to enhance their therapeutic effects.

IMMUNITY: AN ALLY AND OBSTACLE

Oncolytic viruses can modulate the host immune response through various mechanisms. One way is the direct destruction of cancer cells, causing the release of tumor antigens and molecules like high-mobility group box 1 (HMGB1) into the body, which can be recognized by the host immune system. As a result, immune cells like T cells may become activated and begin to hunt down and destroy cancer cells, which may provide long-term protection against tumor recurrence. Oncolytic viruses can be modified to express immune-stimulatory molecules, such as cytokines or co-stimulatory molecules. For instance, some oncolytic viruses are designed to express granulocyte-macrophage colony-stimulating factor (GM-CSF), which can stimulate the production and activation of immune cells like dendritic cells and macrophages that are involved in mounting and maintaining an anti-tumor immune response (Malhotra et al., 2007).

A significant barrier to the clinical development of oncolytic viruses is the immune response, as the immune system may identify the virus as foreign and generate an immune response against it, reducing the treatment's efficacy. Both preexisting immunity and the emergence of an immune response during therapy can lessen effectiveness. The use of immune-modulating medications, genetic modification of viruses to express immune-suppressive molecules, or combination therapy with immune checkpoint inhibitors are some of the ways being investigated by researchers to address these issues.

TYPES OF ONCOLYTIC VIRUSES

DNA VIRUSES

Most extensively studied DNA OVs include adenoviruses and herpes simplex viruses. Adenoviruses, such as ONYX-015, have shown efficacy in clinical trials against various cancers. They exhibit

Oncolytic Viruses

strong tumor-selective replication and can be engineered to express therapeutic genes to enhance their anti-cancer effects. Herpes simplex virus type 1-based oncolytic viruses, like T-VEC, have also demonstrated success in clinical trials, particularly in the treatment of advanced melanoma. DNA oncolytic viruses possess several features and unique characteristics that make them attractive for cancer therapy. Some of them are mentioned as follows.

- Stability and genetic manipulation: DNA-based oncolytic viruses have stable genomes, allowing for efficient genetic manipulation.
- High payload capacity: DNA viruses have a larger capacity to carry and deliver therapeutic genes compared to RNA viruses. This allows for the incorporation of additional therapeutic genes, such as immunomodulators or tumor suppressor genes, into the viral genome for improved virotherapeutic effects.
- Efficient transduction and gene expression: DNA oncolytic viruses efficiently transduce target cells and express their genetic material, ensuring effective replication within tumor cells. This contributes to high viral progeny production and ultimately increased tumor cell killing.
- Enhanced immune activation: DNA viruses possess intrinsic immunogenic properties. Infection by DNA oncolytic viruses triggers the release of pathogen-associated molecular patterns (PAMPs) and danger-associated molecular patterns (DAMPs), leading to the activation of innate and adaptive immune responses against the tumor.
- Long-term transgene expression: DNA oncolytic viruses can sustain long-term transgene expression within the tumor microenvironment. This is particularly advantageous for delivering therapeutic genes that require prolonged expression for their anti-cancer effects.

RNA VIRUSES

RNA OVs such as reoviruses and vesicular stomatitis viruses (VSV) possess inherent tumor-selective properties. Reovirus-based oncolytic viruses, such as Reolysin, have shown efficacy in preclinical and clinical studies against various cancer types. VSV-based oncolytic viruses, such as VSV-GP, have also demonstrated promising results in preclinical models and clinical trials. RNA oncolytic viruses have the potential to revolutionize cancer treatment by exploiting the unique biology of cancer cells and stimulating anti-tumor immune responses. Some common characteristics are as follows:

- Rapid and efficient replication: RNA viruses have high replication rates, enabling them to rapidly amplify within tumor cells. This robust replication leads to the production of a large number of viral progeny, enhancing their oncolytic efficacy.
- RNA as a transgene delivery platform: RNA viruses can be utilized as delivery vectors for therapeutic transgenes. RNA-based oncolytic viruses can carry and express therapeutic genes, such as cytokines, checkpoint inhibitors, or tumor suppressor genes, directly within the tumor microenvironment, augmenting the anti-cancer effects.
- Potential for intratumoral administration: RNA oncolytic viruses can be administered directly into tumor masses, either via direct injection or intralesional infusion. This allows for targeted delivery of the virus to the tumor site, increasing its concentration within the tumor and minimizing systemic exposure.
- Adaptability and ease of genetic manipulation: RNA viruses have relatively simple genomes, making them amenable to genetic manipulation. This allows for the efficient incorporation of modifications, such as enhancing tumor specificity, improving safety profiles, or increasing replication potency.
- Low risk of genomic integration: RNA oncolytic viruses have a low risk of integrating into the host genome, reducing the potential for insertional mutagenesis and genomic instability.

150   Textbook of General Virology

- Immunostimulatory effects: RNA oncolytic viruses possess inherent immunostimulatory properties. They can trigger the activation of the innate immune system, leading to the release of pro-inflammatory cytokines and chemokines. This promotes an anti-tumor immune response, aiding in the elimination of cancer cells.

COMPARISION OF DNA AND RNA ONCOLYTIC VIRUSES

Extensive studies have been conducted on both DNA and RNA oncolytic viruses for their potential as cancer therapies. Although their systems have certain similarities, they also differ from one another. DNA oncolytic viruses, such as adenovirus and herpes simplex virus, are easier to engineer and modify than RNA viruses, which can make them more versatile for use in a variety of cancer types. They may also be less prone to mutation and less likely to cause an immune response, which can improve their efficacy. Single-stranded RNA viruses, such as the reovirus and measles virus, preferentially infect proliferating cells and cause oncolysis by directly lysing the target cells and generating an immunological response. RNA viruses can also cause immunogenic cell death, which triggers an immune response that is anti-tumor, thus having a broader range of targets within cancer cells, which can make them more effective against heterogeneous tumors (Russell, 2002).

According to a study conducted by Lundstrom (2023), RNA viruses are more efficient than DNA viruses. Using a mouse model of melanoma, another study demonstrated that a DNA virus was superior to an RNA virus at triggering immune activation and anti-tumor responses (Bommareddy et al., 2018). It can be suggested that the oncolytic efficacy of viruses can differ based on the targeted tumor, the intended mechanism of action, and the properties of the virus being used. In conclusion, both DNA and RNA oncolytic viruses have shown promising potential in cancer therapies, but further study is required to completely comprehend the variations in their methods of action.

MEASURE OF ONCOLYTIC POTENCY OF ONCOLYTIC VIRUSES

Comparing the oncolytic potency of different viruses is influenced by various factors like the type of cancer being targeted, the route of administration, and the dose and frequency of virus administration. A thorough strategy using many tests and models is required for choosing the most efficient OV in each case. The following provides an explanation of some typical comparative techniques, such as cell viability tests, virus replication assays, and animal tumor models.

- In vitro experiments: In these assays, several oncolytic viruses are used to infect cancer cells, and the amount of virus-induced cell death is then measured. The oncolytic potency of the viruses can be evaluated using a variety of assays, including the plaque assay, the 50% tissue culture infectious dose (TCID50) assay, and the cytopathic effect (CPE) assay.
- In vivo effectiveness studies: These investigations involve administering several oncolytic viruses to cancer-induced animal models (such as mice). Monitoring tumor development or regression, survival rates, and other indicators can be used to gauge the viruses' effectiveness.
- Pharmacokinetics and biodistribution studies: These involve monitoring the amount of the virus over time in the blood and different organs. These tests can reveal details about the virus's capacity to target and infect tumor cells, as well as its rate of elimination from the body.
- Immunological tests: In these tests, the immune response elicited upon OVs administration to cancer cells is measured. The strength and effectiveness of the immune response can be evaluated using a variety of assays, such as checking the concentrations of cytokines and chemokines or examining the activation and growth of immune cells.

Oncolytic Viruses

CURRENT CHALLENGES OF ONCOLYTIC VIRUSES

Despite the potential benefits of oncolytic virus therapy, there are several challenges that need to be addressed to improve the efficacy and safety of this approach.

- Preexisting immunity: Many patients have preexisting immunity to the oncolytic virus, which can reduce the effectiveness of treatment. This can be overcome by using alternate routes of administration, such as intratumoral injection or inhaled delivery, or by modifying the virus to evade immune detection (Russell and Peng, 2007).
- Limited tumor selectivity: Some oncolytic viruses can also infect and replicate in normal cells, which can lead to toxicity and adverse effects. Strategies to improve tumor selectivity include modifying the virus to target specific tumor markers or combining the virus with other cancer therapies that target the tumor microenvironment.
- Drug resistance: Some tumors can develop resistance to oncolytic virus therapy, either by mutating to prevent viral replication or by developing immune evasion mechanisms. Combination therapies that target multiple pathways may be necessary to overcome this resistance (Breitbach et al., 2016).
- Delivery challenges: The delivery of oncolytic viruses to tumors can also be a challenge, especially for deep-seated or metastatic tumors. Novel delivery approaches, such as using nanoparticles or engineering the virus to cross the blood–brain barrier, may improve the efficiency of treatment.
- Regulatory approval: The regulatory approval process for oncolytic viruses can be complex and time-consuming, which can delay their clinical implementation. Streamlining the regulatory process and increasing collaboration between regulatory agencies and researchers may help accelerate the development and approval of oncolytic virus therapies.

CLINICAL TRIALS IN ONCOLYTIC VIRUS THERAPY

Clinical trials with DNA oncolytic viruses have demonstrated promising results, including tumor regression, prolonged survival, and improved quality of life. However, the overall efficacy can vary depending on the tumor type, stage, treatment regimen, and patient characteristics. Adenoviruses are commonly used as DNA oncolytic viruses in clinical trials. Clinical studies have investigated their efficacy in various cancers, including head and neck cancer, prostate cancer, lung cancer, and pancreatic cancer. For example, ONYX-015, an E1B-55kD-deleted adenovirus, has been tested in several clinical trials. Phase III clinical trials demonstrated that the adenovirus-based oncolytic virus significantly improved overall survival in patients with advanced head and neck cancer compared to chemotherapy.

Herpes simplex virus (HSV)-based oncolytic viruses have been studied extensively in clinical trials. Talimogene laherparepvec (T-VEC) is an FDA-approved oncolytic HSV for the treatment of advanced melanoma. Clinical trials have also explored the use of other engineered HSVs in different types of cancers. Phase III clinical trials demonstrated that T-VEC significantly improved durable response rates and overall survival in patients with unresectable metastatic melanoma compared to GM-CSF treatment (Andtbacka et al., 2015). Another OV, Rigvir, is a non-pathogenic ECHO-7 virus that has been used in clinical trials for various types of cancer. Phase II clinical trials demonstrated that Rigvir significantly improved overall survival in patients with melanoma compared to control groups (Doniņa et al., 2015).

A naturally existing oncolytic virus, Maraba virus has been evaluated in clinical trials for various types of cancer. Phase I clinical trials demonstrated that its intravenous delivery was well tolerated and had anti-tumor activity in patients with advanced solid tumors (Pol et al., 2014). Phase I clinical trials demonstrated that the intravenous delivery of the measles virus-based oncolytic virus,

MV-NIS, was well tolerated and had anti-tumor activity in patients with recurrent glioblastoma (Russell et al., 2014).

OVERCOMING BARRIERS TO ONCOLYTIC VIRUS THERAPY

There are several strategies that are being explored to help overcome host immunity against oncolytic viruses during oncolytic virus therapy. Some of these strategies include:

- Using immune-suppressive drugs: Certain drugs, such as corticosteroids, can suppress the immune system and prevent it from attacking the oncolytic virus. These drugs can be given in combination with the oncolytic virus to help prevent the host immune system from recognizing and attacking the virus.
- Modifying the virus: Researchers can modify the oncolytic virus in several ways to help it evade the host immune system. For example, they can engineer the virus to express molecules that suppress the host immune response or to express molecules that promote immune cell infiltration into the tumor microenvironment.
- Altering the route of administration: The route of administration of the oncolytic virus can also play a role in the host immune response. Researchers are exploring different routes of administration, such as intratumoral injection or intravenous administration, to determine which route may be the most effective in avoiding immune detection and enhancing the anti-tumor immune response.

FUTURE PRESPECTIVES

Oncolytic virus research is advancing very rapidly and has a promising future. Currently, research is underway in combination therapy that includes enhancing the anti-tumor immune response and improving overall treatment outcomes by combining oncolytic viruses with other cancer therapies, such as chemotherapy, radiation therapy, and immunotherapy. Creating innovative strategies to specifically target and deliver oncolytic viruses to tumor cells to avoid host immune response and direct delivery inside solid tumors may enhance the efficacy and safety of this therapeutic strategy. For this treatment strategy to advance and receive regulatory approval, it will be crucial to conduct sizable, carefully planned clinical trials to assess the safety and effectiveness of oncolytic viral therapy for diverse forms of cancer.

Personalized treatment by tailoring oncolytic virotherapy can be more effective and have lower side effects if customized to the unique genetic and molecular features of an individual patient's treatment duration. The development of predictive biomarkers that can identify patients who are most likely to benefit from oncolytic virus therapy is an active area of research. Integrating these biomarkers into clinical practice will enable oncologists to select patients who are most likely to respond, sparing others from potential side effects and optimizing resource allocation. By harnessing the power of oncolytic viruses in a personalized manner, the field of cancer medicine can move toward more precise and effective treatments, ultimately leading to improved patient outcomes.

All things considered, oncolytic virotherapy shows great promise as a unique strategy for the treatment of cancer, and ongoing study and development in this area are anticipated to result in substantial advancements in the upcoming years.

REFERENCES

Andtbacka, R. H., Kaufman, H. L., Collichio, F., Amatruda, T., Senzer, N., Chesney, J., ... Coffin, R. S. (2015). Talimogene laherparepvec improves durable response rate in patients with advanced melanoma. *Journal of Clinical Oncology, 33*(25), 2780–2788.

Oncolytic Viruses

Bommareddy, P. K., Shettigar, M., & Kaufman, H. L. (2018). Integrating oncolytic viruses in combination cancer immunotherapy. *Nature Reviews. Immunology, 18*(8), 498–513. https://doi.org/10.1038/s41577-018-0014-6

Breitbach, C. J., Lichty, B. D., & Bell, J. C. (2016). Oncolytic viruses: Therapeutics with an identity crisis. *EBioMedicine, 9*, 31–36.

Doniņa, S., Strēle, I., Proboka, G., Auziņš, J., Alberts, P., Jonsson, B., ... Muceniece, A. (2015). Adapted ECHO-7 virus Rigvir immunotherapy (oncolytic virotherapy) prolongs survival in melanoma patients after surgical excision of the tumour in a retrospective study. *Melanoma Research, 25*(5), 421.

Hou, W., Chen, H., Rojas, J., Sampath, P., & H. Thorne, S. (2014). Oncolytic vaccinia virus demonstrates anti-angiogenic effects mediated by targeting of VEGF. *International Journal of Cancer, 135*(5), 1238–1246.

Hua, L., & Wakimoto, H. (2019). Oncolytic herpes simplex virus therapy for malignant glioma: current approaches to successful clinical application. *Expert Opinion on Biological Therapy, 19*(8), 845–854.

Lundstrom, K. (2023). Self-replicating vehicles based on negative strand RNA viruses. *Cancer Gene Therapy, 30*(6), 771–784.

Malhotra, S., Kim, T., Zager, J., Bennett, J., Ebright, M., D'Angelica, M., & Fong, Y. (2007). Use of an oncolytic virus secreting GM-CSF as combined oncolytic and immunotherapy for treatment of colorectal and hepatic adenocarcinomas. *Surgery, 141*(4), 520–529.

Martuza, R. L., Malick, A., Markert, J. M., Ruffner, K. L., & Coen, D. M. (1991). Experimental therapy of human glioma by means of a genetically engineered virus mutant. *Science, 252*(5007), 854–856.

Patel, P., Young, J. G., Mautner, V., Ashdown, D., Bonney, S., Pineda, R. G., ... James, N. D. (2009). A phase I/II clinical trial in localized prostate cancer of an adenovirus expressing nitroreductase with CB1984. *Molecular Therapy, 17*(7), 1292–1299.

Pol, J. G., Zhang, L., Bridle, B. W., Stephenson, K. B., Rességuier, J., Hanson, S., ... Lichty, B. D. (2014). Maraba virus as a potent oncolytic vaccine vector. *Molecular Therapy, 22*(2), 420–429.

Russell, S. J. (2002). RNA viruses as virotherapy agents. *Cancer Gene Therapy, 9*(12), 961–966.

Russell, S. J., Federspiel, M. J., Peng, K. W., Tong, C., Dingli, D., Morice, W. G., ... Dispenzieri, A. (2014, July). Remission of disseminated cancer after systemic oncolytic virotherapy. In *Mayo Clinic proceedings* (Vol. 89, No. 7, pp. 926–933). Elsevier.

Russell, S. J., & Peng, K. W. (2007). Viruses as anticancer drugs. *Trends in Pharmacological Sciences, 28*(7), 326–333.

19 Emerging and Transboundary Viral Diseases

Nagendra Nath Barman and Lukumoni Buragohain
College of Veterinary Science, Assam Agricultural
University, Guwahati, Assam, India

DEFINITIONS

Rapid increases in the human population have resulted in unprecedented intensification of agriculture, livestock, and other allied farming activities to meet human food demands around the globe. The impact of forest land use brings wild animal species into direct contact with livestock and humans. As a result, many emerging pathogens, including zoonotic ones, are spilled over to livestock and humans (Jagadeesh, 2013). The use of the term *emerging disease* refers to changes in disease dynamics in the population. According to the World Organization for Animal Health (WOAH),

> an emerging disease refers to the changes in the disease dynamics in the population. It is defined as a new infection resulting from the evolution or change of an existing pathogen due to change of host range, vector, pathogenicity of strain or the occurrence of a previously unrecognized infection or disease.

A *re-emerging disease* is considered an already existing disease that either shifts its geographical setting, expands its host range, or significantly increases its prevalence. *Transboundary animal diseases* (TADs) are defined as highly contagious and transmissible epidemic diseases of livestock that can spread rapidly to new areas and regions regardless of national borders, having serious socio-economic as well as public health consequences (FAO, 2021; WOAH, 2021). As per FAO (Food and Agriculture Organization) and WOAH (WOAH) lists (Table 19.1), TADs include African horse sickness, African swine fever, avian influenza, bluetongue, classical swine fever, MERS, Rift Valley fever, sheeppox/goatpox, swine vesicular disease, rinderpest, foot-and-mouth disease, and lumpy skin disease (Lederberg et al., 1992; Daszak et al., 2000). *Exotic diseases* are infectious diseases that are typically not detected or were previously eradicated in a region.

TADs are included in emerging infectious diseases, exotic diseases, and zoonotic diseases. The terms "transboundary animal diseases" and "exotic diseases" are often used synonymously. All transboundary diseases are of exotic origin; however, not all exotic diseases are on the TAD list.

HISTORY

The term "emerging diseases" has been used in scientific literature since the early 1960s (Maurer, 1962). However, the impact of emerging infectious diseases (EIDs) was realized by the scientific community only after the outbreaks of genital herpes and HIV/AIDS in the 1970s (Fleming et al., 1997) and early 1980s, respectively. The significance of EIDs was identified by the Institute of Medicine (IOM) in 1992, thereby providing a guide for their recognition and prevention (IOM, 1992). The establishment of the Program for Monitoring Emerging Diseases (ProMED) contributed greatly to the rapid dissemination of information on infectious disease outbreaks and promoted

154   DOI: 10.1201/9781003369349-19

Emerging and Transboundary Viral Diseases

TABLE 19.1
FOA and WOAH List of Emerging Infectious Diseases and Transboundary Animal Diseases

Viral Disease	Virus Genus	Virus Species	Species Affected
African swine fever (ASF)	*Asfivirus*	*African swine fever virus*	Pigs
Porcine dermatitis nephropathy syndrome	*Circovirus*	*Porcine circovirus 3*	Pigs
Lumpy skin disease (LSD)	*Capripoxvirus*	*Lumpy skin disease virus*	Cattle
Torque teno virus infection	*Iotatorquevirus, Kappatorquevirus*	*Torque tenosus virus*	Pigs
Porcine astrovirus infection	*Mamastrovirus*	*Porcine astrovirus lineage 2* and *4*	Pigs
Porcine reproductive and respiratory syndrome (PRRS)	*Betaarterivirus*	*Porcine reproductive and respiratory syndrome virus*	Pigs
Swine vesicular disease Swine	*Enterovirus*	*Swine vesicular disease virus*	Suids
Vesicular stomatitis	*Vesiculovirus*	*Vesicular stomatitis virus*	Horses, cattle, and suids; rarely sheep and goats
Bluetongue	*Orbivirus*	*Bluetongue virus*	Domestic and wild ruminants; primarily sheep
Sheeppox and goatpox	*Capripoxvirus*	*Sheeppox virus* and *Goatpox virus*	Sheep and goats
African horse sickness	*Orbivirus*	*African horse sickness virus*	Equids; primarily horses
Porcine rotaviral diarrhea	*Rotavirus*	*Rotavirus C*	Pigs
Nipah virus	*Henipavirus*	*Nipah virus*	Pig, horse, domestic and feral cats
Avian influenza	*Influenza A*	*Avian influenza virus*	Domestic poultry; birds and mammals
Middle East respiratory syndrome	*Betacoronavirus*	*Middle East respiratory syndrome coronavirus*	Camels
Rift Valley fever	*Phlebovirus*	*Rift Valley fever virus*	Ruminants
Severe acute respiratory syndrome coronavirus 2 (SARS-CoV-2)	*Betacoronavirus*	Severe acute respiratory syndrome-related coronavirus	Domestic and wild mammalian species
Monkeypox	*Orthopoxvirus*	*Monkeypox virus*	Monkey, gambian pouched rats, squirrels, dormice, non-human primates

communication and collaboration among the international infectious diseases community. Looking back at history, human populations over centuries have experienced major epidemics of infectious diseases. Smallpox carried by explorers from other parts of the world were responsible for 10–15 million deaths in 1520–1521 and ultimately ruined the Aztec civilization. The establishment of germ theory led to enormous progress, notably the development of vaccines and antimicrobials (Bhatia et al., 2012). The Institut Pasteur, established in Paris in 1888, is one of the oldest institutions initiated in research on EIDs. Emile Marchoux in 1896 reported the first outbreak of yellow fever (Augustine, 1909). The laboratory in Saint-Louis studied the epidemiology of malaria and sleeping sickness, and developed vaccines against smallpox, rabies, and the plague. The laboratory was later transferred to Dakar in 1924, presently the World Health Organization (WHO) reference center. In the 1890s, an outbreak of rinderpest south of the River Zambezi killed about 5.2 million cattle, sheep, goats, and oxen, and an unknown number of wild giraffes, buffaloes, and wildebeests. This rinderpest epidemic was the most devastating epidemic to hit southern Africa in the late 19th century. Rift Valley fever (RVF) is a virus from the *Phlebovirus* genus, first identified in the Rift Valley of Kenya in 1931.

LIST OF PATHOGENS

See Table 19.1.

DIAGNOSTIC TECHNIQUES

Emerging infectious diseases/transboundary animal diseases can spread rapidly, causing enormous losses to the health of animals and humans, as well as large costs to society. Despite recent significant advances in the diagnostics of infectious diseases, the need for fast and accurate diagnostic tests of infection in an outbreak situation is obvious to identify the source or epicenter so that appropriate healthcare measures can be quickly instituted. Solid-phase enzyme immunoassays (EIAs) and real-time polymerase chain reaction (PCR) technology, in particular, have revolutionized diagnostic virology. Genotypic or molecular identification methods in diagnostic laboratories are rapidly expanding the ability to detect unusual fastidious pathogen infections.

The risk of a pandemic emerges with EID/TAD pathogenic strains or results from increased virulence, increased transmissibility, or a jump from an animal reservoir. These scenarios often arise from genetic mutation or genetic reassortment (Codit, 2001). Direct or "label-free" sensor-based techniques are suitable for the detection of EID as well as TAD viruses. As a biochemical sensor, immunoassays to detect viral agents still remain quite common (Vemula et al., 2016; Tavakoli et al., 2017). Detection of viruses can specifically be achieved through sandwich enzyme-linked immunosorbent assay (sELISA) with the use of monoclonal antibodies. As a point of care (POC) diagnostic, lateral flow assays (LFAs) are currently being developed for viral detection in a wide range of applications. For a number of reasons, LFAs are attractive diagnostics, especially in resource-limited countries. Quantum dots (QDs) are another emerging visualization reporter for use in LFAs (Hildebrandt et al., 2017). QDs have the ability to be quite sensitive detection agents and can also be easily integrated into multiplexed assays.

PCR is considered one of the gold standard methods for detecting and identifying viral agents. The careful design of target-specific primers and optimization of reaction conditions allow viral agents to be detected at extremely low levels. Loop-mediated isothermal amplification (LAMP) is another diagnostictool that is well-suited for field-based assays for viral detection as it performs at temperatures between 60°C and 65°C. Microarray technology has been used for the simultaneous detection of a large panel of viruses following a multiplex PCR and reverse transcription PCR (RT-PCR) reaction Additionally, next-generation sequencing techniques involve DNA sequencing directly from DNA fragments without the need for cloning in vectors, allowing for the generation of enormous amounts of sequence data at high speed and low cost from a single run. This approach is applied for discovering new viral pathogens associated with EIDs/TADs. Photoluminescence (PL) and fluorescence-based methods likely represent the most widely employed detection methodologies of all sensor technologies (Faltin et al., 2013; Stobiecka & Chalupa, 2015).

RISK FACTORS

Diseases may emerge and re-emerge around the world for a number of interconnected reasons. Factors involved in these outbreaks include:

- Increasing human population puts demand on intensification of farming systems for the relief from hunger and the prevention of famine.

Emerging and Transboundary Viral Diseases

- Increasing demand for animal protein is leading to changes in farming practices (e.g., large commercial farming and backyard farming), animal markets, and natural animal habitats (e.g., encroachment on forests).
- Human and domestic animal encroachment into wildlife habitat and the resulting exposure to wild animals
- Human behavioral changes, including changes in the extent of air travel, ecotourism, hunting, camping, etc.; food preferences (e.g., wild animals and raw milk); and level of compliance with recommended prevention measures.
- Shortfalls in public health infrastructure and policy result from the lack of integration with animal health surveillance, funding in the public health sector, and sustained funding for scientific studies to answer public health questions and build expertise.
- Factors associated with the disease-causing agent include adaptation to new vectors and hosts, mutation and recombination/reassortment in humans and other animals after exposure to multiple pathogens (e.g., foodborne viruses, influenza viruses), and the development of increased virulence.
- The global trade of products, including agricultural products, facilitates the transfer of infectious agents. Both developed and developing countries have processes in place to control the import and export of food and agricultural products to meet sanitary and phytosanitary trade standards as mandated by the World Trade Organzation (WTO). But many times, rules are violated, favoring the spread of EIDs/TADs.
- Wildlife trade is another important and often overlooked factor in potential disease transmission. In the U.S. alone, over 500,000 shipments of wildlife containing 1.4 billion live animals have been imported since 2000.

INCIDENCE AND GLOBAL SPREAD

The incidence of EID events has increased since 1940. The threat of EIDs to global health is increasing with time (Morens et al., 2004; Smolinski et al., 2003; King et al., 2006) and reached a peak in EID events in the 1980s. Increased susceptibility to infection caused the highest proportion of events from 1980 to 1990 (25.5%). The spike in EID events in the 1980s is due largely to the emergence of new diseases associated with the HIV/AIDS pandemic (Lederberg et al., 1992; Smolinski et al., 2003). The majority (60.3%) of EID events are caused by zoonotic pathogens. Furthermore, 71.8% of these zoonotic EID events were caused by pathogens with a wildlife origin (e.g., the emergence of Nipah virus in Perak, Malaysia, and SARS in Guangdong Province, China) and constituted 52.0% of EID events in the most recent decade (1990–2000). This supports the suggestion that zoonotic EIDs represent an increasing and very significant threat to global health (Smolinski et al., 2003; Morens et al., 2004; Weiss & McMichael, 2004). Vector-borne diseases are responsible for 22.8% of EID events and 28.8% in the last decade. This rise corresponds to climate anomalies occurring during the 1990s (Patz et al., 2005). Again, TADs have become a great concern due to the risk to national security on account of their economic significance, zoonotic nature, and ever-growing threat of newer TADs in the future. Among the TADs having zoonotic manifestations, a number of infectious diseases, such as highly pathogenic avian influenza (HPAI), bovine spongiform encephalopathy (BSE), West Nile fever, Rift Valley fever, SARS coronavirus, Hendra virus, Nipah virus, Ebola virus, Zika virus, and Crimean-Congo hemorrhagic fever (CCHF), adversely affecting animal and human health, have been in the news in recent times (Malik & Dhama, 2015).

The impact of both new and re-emerging infectious diseases on human populations is affected by the rate and degree to which they spread across geographical areas, depending on the movement of human hosts or the vectors or reservoirs of infections (Cliff et al., 2000). An increase in travel-associated importations of diseases was anticipated as early as 1933 when commercial air travel was still in its infancy. This has since been dramatically demonstrated by the international airline hub-to-hub pandemic spread of acute hemorrhagic conjunctivitis in 1981 (Morens, 1998), by epidemics of meningococcal meningitis associated with the Hajj, and more recently by the exportation of epidemic SARS (a newly emerging disease) from Guangdong Province, China, to Hong Kong, and from there to Beijing, Hanoi, Singapore, Toronto, and elsewhere (Peiris et al., 2003).

CONSEQUENCES OF EIDS

TADs/EIDs are capable of threatening the global food supply by the direct loss of animal protein and products such as milk, or through production deficits from the loss of animal power, reducing the availability of other animal products such as hides or fibers (Clemmons et al., 2021). For instance, during the BSE outbreak, the annual growth of meat products decreased by 2% in the late 1990s. Pests and animal diseases cause the loss of more than 40% of the global food supply, posing a clear threat to the residual economies of developing countries and the food security of their inhabitants (FAO, 2021). There are significant socio-economic consequences from the cost of control or prevention measures. There is a high likelihood that these diseases can increase poverty and food insecurity, especially in developing nations. Further, emerging infectious diseases account for 26% of annual deaths worldwide. Nearly 30% of 1.49 billion disability-adjusted life years (DALYs) are lost every year to diseases of infectious origin (Taylor et al., 2001; WHO, 2005). The World Bank has estimated that zoonotic disease outbreaks in the past ten years have cost the world more than US$200 billion due to losses in trade, tourism, and tax revenues (Okello, 2011).

One cannot forget that the fight against zoonoses starts by eliminating the pathogen at its animal source. This fact provides veterinary services, veterinarians, farmers, managers of wildlife, and the OIE with a leading role at both national and international levels.

PREVENTION AND CONTROL

Animal health, human health, food production, and the environment are inextricably linked. People depend on livestock and poultry for their livelihoods. The intensification of livestock and poultry farming is increasing. As such, infectious diseases have an ever-increasing importance worldwide, and pathogens do not recognize national boundaries. It is believed that EIDs and TADs will continue to remain an ever-growing threat to animal and human health, the economic sustenance of the world, and global environmental well-being. To mitigate the risks associated with emerging animal diseases, more robust surveillance and control measures need to be put in place, particularly in parts of the developing world where veterinary services and infrastructure remain limited and underresourced. The steps suggested include early detection and early response, preparedness plans, decentralization of government structures, international coordination, understanding of ecology, microbial evolution and viral traffic, disease intelligence, preparedness, and collaboration and cooperation among government agencies. In the future, veterinary or animal medicines are likely to play an increasingly central role in effective disease control. While governments, industry, and regulators around the world have already taken important steps to ensure that veterinary medicines can be delivered quickly and effectively when needed, many challenges remain. As such, this challenge mandates the need for the medical, veterinary, and public health communities to work together as a One Health approach locally and internationally to protect human health, animal health, food safety, and food security (Chen et al., 2022). Using a multisectoral and transdisciplinary approach, EID threats can be better monitored and controlled. Again, the One Health approach to the risk assessment of and response to EIDs focuses on the interconnections between human health, animal health, and the environment. This strategy is essential for the resilience of the EID response system and public health systems. The One Health approach enhances the knowledge of zoonotic diseases and can share information about diseases between animals and humans, with the goal of achieving better health outcomes (Overgaauw et al., 2020; Laing et al., 2021).

Emerging and Transboundary Viral Diseases

SOME OF THE RECENTLY EMERGING AND TRANSBOUNDARY VIRUSES

AFRICAN SWINE FEVER VIRUS

African swine fever (ASF) is a highly fatal viral infectious disease of domestic and wild pigs that manifests as a hemorrhagic fever in affected pigs (Penrith & Vosloo, 2009). ASF has been listed as a priority disease by the OIE, due to its high mortality rate of up to 100%, catastrophic socio-economic importance, trade restrictions, and transboundary potential (OIE, 2020).

History

The first outbreak of ASF was recognized retrospectively in 1907 and described in 1921 in Kenya by Robert Eustace Montgomery. His remarkable publication in 1921 was a paper on "a form of swine fever" (Montgomery, 1921), which today is known as African swine fever. It spread to several sub-Saharan African countries and remained on the African continent until 1957. Slowly, African swine fever reached Europe in 1957 when it was detected on a Portugal pig farm and tehn spread to Europe, Cuba, Brazil, and the Dominican Republic. It slowly spread to China in 2018 and was found in over a dozen other Asian countries, including India, in 2020 (Bora et al., 2020).

Etiology and Natural Reservoirs of ASF

African swine fever is caused by the *African swine fever virus* (ASFV), the only species belonging to the genus *Asfivirus* of the family *Asfarviridae* (Alonso et al., 2018). The virion has a very complex structure and an overall diameter of 175–215 nm. The virion consists of a nucleoprotein core (70–100 nm in diameter), a core shell surrounded by an internal lipid layer, an icosahedral capsid with 1892–2172 capsomers, and a dispensable lipid envelope (Alonso et al., 2018; Salas & Andrés, 2013). The linear dsDNA molecule encodes 54 different structural proteins, in addition to a number of non-structural proteins. To date, 24 distinct genotypes of ASFV have been described (Reis et al., 2017). ASF can lead to high mortalities in domestic pigs and wild boar. Warthogs and bush pigs and soft ticks of the genus *Ornithodoros* serve as natural reservoirs in ASFV transmission (Costard et al., 2013).

Replication Dynamics

African swine fever virus propagates in the mononuclear-phagocytic cells through clathrin-mediated endocytosis (Sanchez et al., 2012). Upon entry, viral DNA is released through the endosomal pathway. Replication and assembly of virions take place in special virus factories close to the nucleus. Mature virus particles are released from the infected cells by budding (Almazan et al., 1993).

Geographical Distribution

African swine fever remained endemic in most of Africa prior to the 1950s. ASF reached Europe in 1957 and was detected on a Portugal pig farm. In 1960, the disease rapidly spread to Spain and France and also expanded into Cuba, Brazil, and the Dominican Republic in 1978. The disease re-entered the European continent in 2007 and spread across 20 different countries of the European continent (Sánchez-Vizcaíno et al., 2015). More recently, the virus spread to Western Europe in Belgium in 2018 in wild boar. In August 2018, ASFV was reported for the first time in Asia, and by the end of September 2019, ASFV was detected in Cambodia, Laos, Thailand, Vietnam, Mongolia, the Philippines, North Korea, South Korea, Myanmar, Indonesia, Timor-Leste, Papua New Guinea, and in 2020 in India (OIE WAHIS African Swine Fever (ASF) Report: May 15–28, 2020; www.oie.int).

Risk Factors

The main risk factors for ASF spread are grouped into seven categories: poor biosecurity, swill feeding and slaughtering on farm, trading of pigs and products, human activity factors and farm management, sociocultural risk factors, ASF in wild boars, and ticks and other blood-feeding arthropods (Bellini et al., 2021).

Pathogenesis

ASFV is highly contagious, and the infective virus invades through the tonsils and respiratory tract and replicates in the lymphoid tissues of the nasopharynx prior to the development of viremia that occurs within 48–72 hours of infection (Plowright et al., 1994). Primary replication takes place in the monocytes and macrophages of the lymph nodes (Sanchez et al., 2012). The virus spreads via the blood route or via the lymphatic route. As the ASFV spreads to different organs, such as the lymph nodes, bone marrow, spleen, kidneys, lungs, and liver, and secondary replication and the characteristic hemorrhagic lesions develop. The incubation period of the disease ranges from 3 to 19 days. The clinical signs vary depending on the virulence of the ASFV strain. The highly virulent strains result in 90–100% mortality, moderately virulent strains with 70–80% mortality in young pigs and 20–40% mortality in adults, and low virulent strains with 10–30% mortality. Pigs infected with ASFV may develop a peracute or acute form, subacute form, and chronic form, depending on the strain of ASFV infection (Blome et al., 2013).

Isolation

ASFV can be isolated in primary porcine cells (pig leukocyte cells, porcine alveolar macrophages, bone marrow cultures, porcine blood monocytes), monkey-derived established cell lines (Vero, COS-1 cells), and porcine established cell lines (IPAM, ZMAC, WSL cells).

Diagnosis

Appropriate samples for laboratory testing are whole blood, serum from clinically affected animals, and tissues, mainly spleen, lymph nodes, bone marrow, lung, tonsil, and kidney from dead carcasses. Non-invasive sampling strategies include the use of bait (ropes) for the collection of oral fluid in the wild boar surveillance programs. To confirm and differentiate, laboratory assays include virus isolation on porcine macrophages, (real-time) PCR, fluorescent antibody test (FAT), ELISA, immunoblotting, and indirect immunostaining techniques (OIE, 2019). Currently, PCR is considered the gold standard test for early detection of the disease.

Prevention and Control

The control of ASF is a very important task as there is no safe, effective vaccine for ASF. Therefore, control efforts of ASF focus on the elimination of contaminants and the blocking of transmission routes, which will remain a long-term effort and require the involvement of all parties in the pork food industry. Control efforts include timely and reliable diagnosis, stamping out of infected herds, the establishment of restriction zones, movement restrictions, tracing of possible contacts, and implementation of strict biosecurity measures under a strict legal framework (Blome et al., 2013).

LUMPY SKIN DISEASE

Lumpy skin disease is an infectious vector-borne viral disease that affects cattle and Asian water buffalo and is caused by the *lumpy skin disease virus* (LSDV) of the genus *Capripoxvirus*, family *Poxviridae*. The OIE has placed this transboundary disease on the notifiable disease list due to its significant economic losses, the potential for rapid spread, and international trade embargoes (Abdulqa et al., 2016).

History

The first clinical manifestations of lumpy skin disease (LSD) were recorded as skin nodules in 1929 in Zambia (Northern Rhodesia). A similar condition named "Ngamiland cattle disease" occurred in Botswana in October 1943. Between 1943 and 1945, the disease spread to other African countries.

Emerging and Transboundary Viral Diseases

Skin lesions of affected cattle inoculated into healthy cattle resulted in similar nodular lumps in the skin of cattle and thus was named lumpy skin disease (Bhanuprakash et al., 2006). The disease progressively spread northward, and it is now found across the entire African continent, and it later extended to Algeria, Europe, and Asian countries.

Virus Properties and Host Range

LSDV belongs to the *Capripoxvirus* genus of the sub-family *Chordopoxvirinae* of the family *Poxviridae* (King et al., 2012). LSDV is a brick-shaped enveloped virus, 320 × 260 nm in size, with double-stranded DNA having complex symmetry, and the virus replicates in the cytoplasm of the infected cell. The LSDV genome is 151 kbp large; it expresses 30 structural and non-structural genes sharing 97% nucleotide homology with the sheeppox and goatpox virus (Tulman et al., 2002). The virus is stable in ambient environmental conditions for long periods. It can remain unaffected in desiccated skin crusts for 35 days, in necrotic nodules for 33 days, and in air-dried hides for at least 18 days. The virus is inactivated quickly in sunlight and lipid detergents, but the virus can persist for many months in dark premises. The virus is inactivated at 55°C for 2 h and 65°C for 30 minutes. It is susceptible to highly alkaline or acidic pH but can survive at pH 6.6–8.6 for 5 days at 37°C, and virus titers remain unaffected. The virus is susceptible to ether (20%), chloroform, formalin (1%), phenol (2% for 15 minutes), sodium hypochlorite (2–3%), iodine compounds (1:33 dilution), and quaternary ammonium compounds (0.5%) (OIE, 2013). LSDV is very stable and can be recovered even after 10 years from skin nodules kept at −80°C and after 6 months from infected tissue culture fluid kept at 4°C (Mulatu & Feyisa, 2018). Susceptible hosts irrespective of age group are cattle and buffalo (Al-Salihi, 2014). Goats and sheep may be experimentally infected. Wild animals are generally resistant to infection, but several wildlife species were found seropositive to LSDV. The role of wildlife in the transmission and maintenance of LSDV needs further study. Humans are also resistant to the virus (OIE, 2013).

Replication Dynamics

All pox viruses exclusively replicate in the cytoplasm of the infected cell. To perform replication of DNA in the cytoplasm, all essential viral enzymes and factors required for the transcription of the early subset of genes are packaged in the core of infectious virus particles (Moss, 2007). The virus core is then released into the cytoplasm, and transcription is characterized by a cascade as "early," "intermediate," and "late" genes. Progeny DNA serves as the template for the transcription of inter-mediate- and late-stage genes (Yang et al., 2011). Replication and assembly occur in discrete sites within the cytoplasm (called viroplasm), and virions are released by budding (enveloped virions), by exocytosis, or by cell lysis (non-enveloped virions).

Geographical Distribution

LSD is prevalent in most African countries, notably in the sub-Saharan area (Al-Salihi, 2014; Rao et al., 2000). It extended to South-East Europe, the Balkans, the Caucasus, Russia, and Kazakhstan after 2012 (OIE, 2018). The last outbreak reported in Kazakhstan was in 2016. A similar disease outbreak was reported in Bangladesh in July and September 2019. In India, the first outbreak of the disease was reported in Odisha state in August 2019 of high humidity (Sudhakar et al., 2020).

Risk Factors

LSDV is mechanically transmitted by a variety of hematophagous arthropod vectors (*Rhipicephalus, Amblyomma, Aedes aegypti, Stomoxys*). Warm and humid climatic conditions favor the propaga-tion of vector populations. As such, LSD cases can be seen after seasonal rains, the introduction of new animals to a herd (Ince et al., 2016), and through semen (Annandale et al., 2013). Moreover, wind direction and the strength of the wind can contribute to the spread of the virus (Rouby & Aboulsoud, 2016).

Pathogenesis

Lumpy skin disease is a systemic infection with cell-associated viremia. The virus has a distinct tropism for keratinocytes. The epidermal microvesicles coalesce into large vesicles that quickly ulcerate. Nodular proliferation occurs in the lungs and lymphadenopathy. Blood monocytes are important in spreading the virus to secondary sites of infection (Constable et al., 2017). Various studies showed that intracellular replication of the virus occurs in fibroblasts, macrophages, pericytes, and endothelial cells, leading to vasculitis and lymphangitis in affected tissues (Coetzer, 2004). Affected animals clear the infection, and no carrier state is known for LSDV yet (Tuppurainen et al., 2017).

Isolation

Virus isolation is critical, but the virus is utilized in a number of primary cells or cell lines of bovine, ovine, or caprine origin. However, primary bovine dermis and PLT cells are considered to be the most susceptible. The isolated virus was adapted in Vero cell lines (Kumar et al., 2021). The virus may also grow on the chorioallantoic membrane of embryonated chicken eggs and African green monkey kidney (Vero) cells (OIE, 2018). LSDV induces a specific cytopathic effect (CPE) and intracytoplasmic inclusion bodies in cell culture and causes syncytia and intranuclear inclusion bodies in cell culture (Abdulqa et al., 2016; Kumar et al., 2021).

Diagnosis

Lumpy skin disease can be primarily diagnosed by clinical symptoms. However, confirmatory diagnosis can be made by conventional PCR (Zheng et al., 2007) or real-time PCR techniques (Balinsky et al., 2008). For differentiating virulent LSDV from the vaccine strain, restriction fragment length polymorphism (RFLP) has also been used (Menasherow et al., 2014).

Prevention and Control

LSD treatment is only symptomatic, with antimicrobial therapy used to prevent subsequent bacterial infections (Molla et al., 2017). Because movement restrictions and the removal of affected animals are typically ineffective, vaccination is the only practical and economically viable strategy for controlling the spread of the disease and improving cattle productivity in endemic areas (OIE, 2018; Tuppurainen, 2017). Vaccinating animals every year might keep LSD under control (Thomas, 2002). Commercially available vaccines are live attenuated homologous (Neethling LSDV strain) and heterologous (*Sheeppox virus* and *Goatpox virus*) vaccines. However, heterologous vaccines may not provide complete immunity (Namazi et al., 2021).

Nipah Virus

Nipah virus (NiV) is an emerging zoonotic virus that can cause mild to fatal neurological and respiratory illnesses in humans and animals (Chua et al., 2000; Aditi & Shariff, 2019; Singh et al., 2019; Sharma et al., 2019). It causes high morbidity and mortality in humans and it is a global public health threat as there is no approved commercial vaccine for humans. Although NiV emerged 25 years ago, outbreaks are still recorded in India and Bangladesh. Therefore, special attention is necessary to tackle this virus in such prone areas.

Virus Classification and Its Genome Characteristics

Nipah virus is a single-stranded RNA virus belonging to the genus *Henipavirus* of the family *Paramyxoviridae* (Ksiazek et al., 2011; Soman Pillai et al., 2020). NiV is an enveloped virus exhibiting a pleomorphic shape with a size from 40 to 1900 nm (Ang et al., 2018; Skowron et al., 2022). The RNA genome of NiV is non-segmented and negative sense (3′ to 5′), which is about 18.2 Kb in length and encodes six structural proteins, viz., nucleocapsid (N), phosphoprotein (P), matrix protein (M), fusion protein (F), glycoprotein (G), and RNA polymerase (L) (Soman Pillai et al., 2020;

Emerging and Transboundary Viral Diseases

Devnath et al., 2022). The N, P, and L proteins form the ribonucleoprotein complex by attaching to the RNA genome of NiV.

Host and Transmission

Natural reservoirs of NiV are fruit bats of the family *Pteropodidae* and genus *Pteropus* (flying foxes) (Bruno et al., 2022). Although fruit bats are asymptomatic, they act as carriers and disseminate the virus to other susceptible animal hosts (pig, horse, domestic and feral cats) and humans by shedding the virus in saliva, urine, or feces or through contaminated palm fruits (Soman Pillai et al., 2020; Bruno et al., 2022). Dogs and cats can also be infected with NiV (Bruno et al., 2022). The transmission pattern of NiV can be either intraspecies (e.g., human to human) or interspecies (e.g., pig to human). Humans are generally infected when there is direct contact with infected bats, pigs, or horses, or contaminated biological matrices of infected animals such as urine, saliva, blood, and feces. Human-to-human transmission occurs when there is close contact with an infected person or their body fluids, including nasal discharge or respiratory droplets (Soman Pillai et al., 2020; Bruno et al., 2022). Humans also get infected by ingestion of food products (undercooked meat) or fruits (palm fruits or sap) contaminated by infected animals (Soman Pillai et al., 2020; Skowron et al., 2022). There are also reports of NiV infection in persons who have climbed palm trees where fruit bats usually roost (Ang et al., 2018).

Epidemiology

The major cases of NiV infection are reported from Asian countries, particularly from Malaysia, Bangladesh, and India (Soman Pillai et al., 2020). However, an episode of NiV infection was also reported in the Philippines in 2014 (Ching et al., 2015; Skowron et al., 2022). Although active cases were reported from Malaysia, Bangladesh, India, and the Philippines, there is evidence of the presence of antibodies in other countries like Thailand, Cambodia, Ghana, and Madagascar (Soman Pillai et al., 2020). The first Nipah virus infection cases were reported from pig farmers in September 1998 from Ipoh city of Perak, Malaysia (Chua, 2003). This was followed by several outbreaks in different states of Malaysia up to mid-1999 (Chua, 2003). For the first time, NiV was isolated from the cerebrospinal fluid (CSF) of a patient with encephalitis in March 1999 from Sungai Nipah village, giving the official name Nipah virus (Chua, 2003). Bangladesh recorded the first outbreak in April 2001 from Meherpur district (Hsu et al., 2004). From 2001 to early 2015, there were annual outbreaks, particularly in the winter season. In early 2001, India also reported the first outbreak from the Siliguri area of West Bengal (Chadha et al., 2006). The second outbreak was also reported from West Bengal in 2007 but from the district of Nadia (Kulkarni et al., 2013; Soman Pillai et al., 2020). Subsequently, in 2018 and 2021, confirmed cases were reported from Kerala state of India (Arunkumar et al., 2019; Skowron et al., 2022).

Diagnosis

Early and confirmatory diagnosis of NiV infection is essential to pursue necessary prevention and control. Although early diagnosis is challenging due to common symptoms that coincide with many other viral diseases, early diagnosis can increase the survival chance of NiV-infected persons. For the diagnosis of NiV infection, common diagnostic platforms such as virus isolation, immunoassays, and molecular assays are employed. Human clinical specimens that are commonly used for NiV detection include blood, cerebrospinal fluid, urine, nasal and throat swabs, and animal samples, which include blood, lung, spleen, and kidneys (Daniels et al., 2001). Molecular assays like conventional RT-PCR, nested RT-PCR, and real-time RT-PCR are used to detect NiV nucleic acid in clinical samples because molecular assays are very sensitive and specific (Soman Pillai et al., 2020; Skowron et al., 2022). On the molecular platform, the RT-LAMP assay has also been used to detect NiV genome (Ma et al., 2019). Among immunoassays, indirect ELISA is used to detect IgG and IgM antibodies. Other forms of ELISA, such as capture and sandwich ELISA, have been also developed for the diagnosis of NiV infection (Singh et al., 2019). Immunohistochemistry can also be employed to detect the NiV antigen in formalin-fixed tissue samples such as spleen, central nervous system, lung, kidney, lymph nodes, and heart (Bruno et al., 2022). Isolation of NiV in cell culture, particularly in the Vero cell line, is the most confirmatory diagnosis; however, it requires a BSL-4 facility.

Prevention and Control

There is no magical treatment or effective commercial vaccine against NiV for use in humans. Therefore, priority must be given to formulating strong and effective preventive and control measures. Several vaccines are in different developmental stages. A subunit vaccine based on recombinant soluble G glycoprotein (sG) of Hendra virus provides cross-protection against NiV. The only vaccine approved for use in horse is Equivac (Skowron et al., 2022). The common preventive measures that are to be adopted in areas where NiV outbreaks have occurred or areas that are prone to catch the infection (CDC, 2022a; Skowron et al., 2022) include the following:

a) Avoid direct contact with fruit bats and their secretions or excretions.
b) People associated with pig farming must take special care in handling the pigs and should use personal protective equipment (PPE) during the slaughter and disposal of animals.
c) Persons dealing with horses must also avoid direct contact with sick horses and their body fluids.
d) Fruits must be taken only after proper washing, and avoid eating or drinking fruit and fruit juices (palm sap) that could be contaminated by fruit bats.
e) Fruit trees that attract bats should not be planted near pig farms.
f) Avoid direct contact with NiV-positive persons, and health workers must use PPE in dealing with such patients.
g) One should practice proper handwashing with soap or detergent, sanitization with 70% alcohol, and disinfection with physical or chemical methods to be followed wherever necessary.
h) There should be awareness in all forms to inform the public about the signs and symptoms of NiV infection, the risk factors associated with the disease, and preventive measures that should be taken in an outbreak.
i) Regular disease surveillance and monitoring are essential to predicting disease outbreaks or providing early warning about diseases, as well as identifying new environmental factors, new emerging strains of the virus, or new potential reservoir hosts.

SEVERE ACUTE RESPIRATORY SYNDROME CORONAVIRUS 2 (SARS-CoV-2)

One of the terrible pandemics that humanity has witnessed is the current COVID-19 pandemic. It is caused by coronavirus that is suspected to have originated from bat coronavirus and can infect a wide range of mammalian hosts. The pandemic has taken millions of lives and human progress has been reverted by many years. The world was static for two years. The COVID-19 pandemic started in December 2019 in China, and within a short period, it disseminated to all corners of the globe. Today, fortunately, we have vaccines to prevent the disease.

Virus Classification and Its Genome Characteristics

Severe acute respiratory syndrome coronavirus 2 (SARS-CoV-2) belongs to the *Coronaviridae* family, subfamily *Orthocoronavirinae* and genus *Betacoronavirus* (betaCoV) (BeigParikhani et al., 2021). Like other coronaviruses, SARS-CoV-2 is an enveloped, single-stranded positive-sense RNA virus. The genome of SARS-CoV-2 is about 29.9 kb, non-segmented, and arranged linearly as 5′-leader-UTR-replicase-structural genes-3′ UTR-poly(A) tail (Dhama et al., 2020). The genome of SARS-CoV-2 contains six major open reading frames (ORFs) in order from 5′ to 3′: replicase (overlapping ORF1a/ORF1b), spike (S), envelope (E), membrane (M), and nucleocapsid (N) (BeigParikhani et al., 2021; Hu et al., 2021). Additionally, there are eight other ORFs interspersed

Emerging and Transboundary Viral Diseases

between the structural genes that encode accessory proteins: 3a, 3b, p6, 7a, 7b, 8b, 9b, and orf14 (BeigParikhani et al., 2021; Hu et al., 2021). The ORF1a and ORF1b occupy the major portion (2/3) of the 5′ genome and encode 15 non-structural proteins (Nsps), which are Nsp1 to Nsp10 by ORF1a, and Nsp12 to Nsp16 by ORF1b. The 3′ end of the genome encodes four structural proteins: S, E, M and N (BeigParikhani et al., 2021; Hu et al., 2021).

Host and Transmission Pattern

The SARS-CoV-2 virus can infect a wide range of mammalian species, including humans, and domestic and wild animals (Tan et al., 2022). The origin of SARS-CoV-2 is still debated; however, several studies have indicated that the most probable ecological reservoir is bats (Zhao et al., 2020; Tan et al., 2022). Besides humans, SARS-CoV-2 has been isolated from minks, domestic cats and dogs, golden hamsters, white-tailed deer, big cats, and many others (Tan et al., 2022).

One of the theories says that SARS-CoV-2 originated from bat novel coronavirus (bat-nCoV), and from bats, it was transmitted to an intermediary wild animal host where it mutated and evolved, and through the wild animal business, it was transmitted to humans and domestic animals (Zhao et al., 2020). Human-to-human transmission can be through direct or indirect contact. The most potent transmission route is droplet transmission (Han et al., 2020; Wu et al., 2021). An infected patient can spread the virus by coughing, sneezing, breathing, talking, or singing (Han et al., 2020; Wu et al., 2021). Fomite is another compounding route of transmission in humans (Guo et al., 2020; Lei et al., 2020; Ong et al., 2020; Xie et al., 2020). Fomite transmission is an indirect mode of transmission in which an individual is infected through touching infected surfaces or objects. Nosocomial transmission is also one of the potent routes of infection (Wang et al., 2020a; Wu et al., 2021).

Epidemiology

The COVID-19 pandemic started as a pneumonia of unknown origin in Wuhan City, China, and the first cases of COVID-19 in human patients from Wuhan were confirmed in December 2019 (Park, 2020; Yu et al., 2020). Initially, the causative agent was designated as novel coronavirus (2019-nCoV); however, the World Health Organization (WHO) declared the official name of the disease as coronavirus disease 2019 (COVID-19) on 11 February 2020 (Attia et al., 2021). Later, the International Committee on Taxonomy of Viruses (ICTV) named the causative virus as SARS-CoV-2 (Yu et al., 2020). On 30 January 2020, the WHO declared the disease a public health emergency of international concern (PHEIC), and by 11 March 2020 the disease had spread to at least 114 countries with more than 4,000 deaths (Park, 2020; Yu et al., 2020). Thus, on the same day2020), WHO officially declared COVID-19 a pandemic (Park, 2020). By January 2020, the disease had spread to most of the cities of China and many Asian countries (Attia et al., 2021). By February–April 2020, the disease had been reported on all continents, except Antarctica. As of this writing, outbreaks were ongoing in different locations of the globe. As of 10:11 am CEST, 12 April 2023, 762,791,152 cases had been confirmed with 6,897,025 deaths across the globe (https://covid19.who .int/).

Diagnosis

The most popular, sensitive, and reliable molecular diagnostic technique that is used for detection of SARS-CoV-2 nucleic acid is real-time PCR. There are several real-time based RT-PCR commercial kits that target E, N, S and ORF1b genes of SARS-CoV-2 that are being used to diagnose SARS-CoV-2 infection (BeigParikhani et al., 2021; Hu et al., 2021). Besides real-time PCR, other molecular techniques used to detect SARS-CoV-2 nucleic acid are conventional RT-PCR and RT-LAMP (Zhai et al., 2020; BeigParikhani et al., 2021). The common samples used for molecular detections are nasopharyngeal swabs, throat swabs, sputum, and bronchial fluid (Jin et al., 2020; BeigParikhani et al., 2021; Cascella et al., 2023). Detection of SARS-CoV-2 antigen by lateral flow assay (immunochromatic strip) is another method that was well accepted during the pandemic and approved by many national and international agencies (BeigParikhani et al., 2021). Such detection kits are low cost, and results can be available within 10–15 minutes. Serological assays have been employed to detect antibodies against the SARS-CoV-2 virus, particularly ELISA and neutralization assays. However, antibody detection tests are not directly used for diagnosis, but rather for epidemiological studies to identify asymptomatic patients, the immunological status of recovered individuals, and vaccine efficiency (Long et al., 2020; BeigParikhani et al., 2021). Other valuable diagnostic procedures that assist in diagnosis and monitoring of COVID-19 positive patients include imaging modalities such as chest computed tomography (CT), chest X-ray, and lung ultrasound (Cascella et al., 2023).

Prevention and Control

The most convincing method to prevent COVID-19 is by vaccination. Within a period of two years of the pandemic, a good number of vaccine candidates had been approved nationally or internationally for emergency use. The vaccine candidates were developed using different platforms such as mRNA vaccines, viral vector vaccines, killed vaccines, and DNA vaccines. Each vaccine has its own dose and immunization schedule. Commercially available and approved vaccines include Covaxin, Sputnik V, CoronaVac, Moderna COVID-19 (mRNA-1273), Covishield, and ZyCoV-D. As of 10 April 2023, a total of 13,340,275,493 doses of vaccine have been administered across the world (https://covid19.who.int/).

Some of the common preventive measures that are to be adopted are as follows:

a) Maintain social distance (1 meter from COVID-19 suspected or positive persons).
b) Washing of hands regularly or the use of alcohol-based sanitizer.
c) Use of face masks in crowded areas or in hospitals.
d) Health workers must use PPE while treating patients.
e) Isolation of positive patients.
f) Quarantine of individuals who may have contracted the disease.

MONKEYPOX

Monkeypox virus emerged in 1958, but human cases were detected only in the 1970s. Monkeypox (MPX) is an infectious zoonotic disease and exhibits similar clinical manifestations of smallpox but with less severity. Initially, the disease was endemic to African countries; gradually, it spread to other continents and became a global threat in 2022 due to a sudden rise in positive cases. Although there is no vaccine against MPX, the smallpox vaccine helps in reducing the disease incidence manifold.

Virus Classification and Its Genome Characteristics

The monkeypox virus is a double-stranded DNA virus belonging to the family *Poxviridae* and genus *Orthopoxvirus* (McCollum & Damon, 2014). The MPX virus is closely related to smallpox, but the severity is less (Singhal et al., 2022). The name "monkeypox" was coined based on its first isolation from cynomolgus monkeys in 1958 in Denmark (Magnus et al., 1959). The MPX virus is an enveloped virus that exhibits brick-shaped ovoid structure and measures about 200×250 nm (Harapan et al., 2022; Rajsri & Rao, 2022). The MPX virus has a linear dsDNA genome of about 197 kbp and encodes approximately 190 ORFs (Kugelman et al., 2014). The genome is divided into three parts: a core region, a right arm, and a left arm (Wang et al., 2022). Its genome has covalently closed hairpin ends and contains inverted terminal repeats of 10 kbp at both ends (Harapan et al., 2022). The central region of the genome is conserved and contains housekeeping genes responsible for viral replication, transcription, and viron assembly, whereas genes present in the terminal region are involved in pathogenesis.

Emerging and Transboundary Viral Diseases

Host and Transmission Pattern

Susceptible hosts of MPX virus include humans, monkeys, Gambian pouched rats, squirrels, dormice, non-human primates, and several other species (Singhal et al., 2022). Transmission of the MPX virus can occur human to human or animal to human, as it is a zoonotic virus (McCollum & Damon, 2014). In the animal-to-human transmission process, humans are infected due to direct contact with body fluids, blood, mucosal or cutaneous lesions of infected animals, or due to bites or scratches from infected animals (Singhal et al., 2022; WHO, 2022). Transmission may also happen due to the consumption of undercooked meat of infected animals (Singhal et al., 2022; WHO, 2022). Transmission between humans can result from direct contact with infected individuals through respiratory secretions, skin lesions, or respiratory droplets (Singhal et al., 2022; WHO, 2022). Indirect means of contact could be through contaminated fomites (Singhal et al., 2022; WHO, 2022). Transmission via placenta from mother to fetus or newborn is also possible (Mbala et al., 2017).

Epidemiology

Monkeypox was first identified in humans during the 1970s in several African countries and became endemic to many regions of Africa (Lum et al., 2022). For about more than 20 years, the virus was confined to African countries where cases rose from a few to several thousands. In 2003, the first human case outside Africa was reported from the United States (Lum et al., 2022; Singhal et al., 2022). Then between 2018 and 2021, cases were reported in Singapore, the UK, and Israel (Lum et al., 2022; Singhal et al., 2022). In 2022, suddenly there was a drastic increase in cases from different countries. In July 2022, a few cases were reported from India as well. Looking into the increasing rate of incidence in different parts of the world, on 23 July 2022 the WHO declared MPX a public health emergency of international concern (Lum et al., 2022). As of 1 February 2023, a total of 86,956 cases have been confirmed from 110 countries with 119 deaths (CDC, 2022b).

Diagnosis

Suitable clinical specimens required for the laboratory diagnosis of MPX are skin lesion exudate or scabs, lesion crusts, roofs from pustules and vesicles, and dry crusts (Saijo et al., 2008; WHO, 2022). The most reliable and sensitive test to detect the MPX virus is real-time PCR. Other molecular tests include PCR and isothermal amplification such as recombinase polymerase amplification (RPA) and LAMP (Singhal et al., 2022). Other confirmatory tests that are done to detect the MPX virus are virus isolation (in BSL-3 laboratories), electron microscopy, and immunohistochemistry (Petersen et al., 2019; Harapan et al., 2022). Monkeypox virus-specific IGM and IgG can be detected by ELISA for epidemiological investigation, late clinical manifestation cases, chronic cases, and retrospective diagnosis (Titanji et al., 2022). The point-of-care diagnostic kit Orthopox BioThreat Alert® can be used to detect pox virus antigen in field conditions; however, the test is not specific to MPX (Singhal et al., 2022).

Prevention and Control

Adaptation of proper preventive and control measures can inhibit the transmission of the MPX virus. People vaccinated with the smallpox vaccine are less infected than the unvaccinated; thus, the vaccine against the smallpox virus can be used to prevent MPX virus infection. Therefore, people in endemic areas must be vaccinated with the smallpox vaccine. Two vaccines, viz., JYNNEOS and ACAM2000, approved by the US Food and Drug Administration against smallpox, are used to prevent monkeypox. JYNNEOS (Imvanex, Imvamune) contains a non-replicating modified vaccinia virus (Ankara strain) that is given subcutaneously in doses at 28-day intervals (El Eid et al., 2022; Rajsri & Rao, 2022; Singhal et al., 2022). Besides vaccination, other preventive measures include avoiding direct or indirect contact with persons having suspected skin lesions, rapid and early detection and isolation of positive patients, avoiding sharing of materials like clothing with suspected or MPX positive individuals, use of PPE by healthcare workers, quarantining of persons for at least for six weeks from the date of last exposure, raising awareness in rural areas where the disease is endemic, and avoiding contact with susceptible and carrier animal hosts.

REFERENCES

Abdulqa, H. Y., Rahman, H. S., Dyary, H. O., & Othman, H. H. (2016). Lumpy skin disease. *Reproductive Immunology: Open Access* 1, 6.

Aditi & Shariff, M. (2019). Nipah virus infection: A review. *Epidemiology and Infection*, 147, e95. https://doi.org/10.1017/S0950268819000086

Almazan, F., Rodriguez, J. M., Angulo, A., Vinuela, E., & Rodriguez, J. F. (1993). Transcriptional mapping of a late gene coding for the p12 attachment protein of African swine fever virus. *Journal of Virology* 67, 553–556

Al-Salihi, K. (2014). Lumpy skin disease: Review of the literature. *Mirror of Research in Veterinary Sciences and Animals* 3(3), 6–23.

Ang, B. S. P., Lim, T. C. C., & Wang, L. (2018). Nipah virus infection. *Journal of Clinical Microbiology* 56(6), e01875–17. https://doi.org/10.1128/JCM.01875–17

Annandale, C., Cornelius, H., Holm, D.E., Ebersohn, K., & Venter, E. H. (2013). Seminal transmission of lumpy skin disease virus in heifers. *Transboundary and Emerging Diseases* 61, 443–448.

Arunkumar, G., Chandni, R., Mourya, D. T., Singh, S. K., Sadanandan, R., Sudan, P., Bhargava, B., & Nipah Investigators People and Health Study Group (2019). Outbreak investigation of nipah virus disease in Kerala, India, 2018. *The Journal of Infectious Diseases* 219(12), 1867–1878. https://doi.org/10.1093/infdis/jiy612

Attia, Y. A., El-Saadony, M. T., Swelum, A. A., Qattan, S. Y. A., Al-Qurashi, A. D., Asiry, K. A., Shafi, M. E., Elbestawy, A. R., Gado, A. R., Khafaga, A. F., Hussein, E. O. S., Ba-Awadh, H., Tiwari, R., Dhama, K., Alhussaini, B., Alyileili, S. R., El-Tarabily, K. A., & Abd El-Hack, M. E. (2021). COVID-19: Pathogenesis, advances in treatment and vaccine development and environmental impact-an updated review. *Environmental Science and Pollution Research International* 28(18), 22241–22264. https://doi.org/10.1007/s11356-021-13018-1

Augustine, G. (1909). *History of yellow fever.* Searcy and Pfaff.

Balinsky, C. A., Delhon, G., Smoliga, G., Prarat, M., French, R. A., Geary, S. J., Rock, D. L., & Rodriguez, L. L. (2008). Rapid preclinical detection of Sheeppox virus by a real-time PCR assay. *Journal of Clinical Microbiology*, 46, 438–442. https://doi.org/10.1128/JCM.01953-07

BeigParikhani, A., Bazaz, M., Bamehr, H., Fereshteh, S., Amiri, S., Salehi-Vaziri, M., Arashkia, A., & Azadmanesh, K. (2021). The inclusive review on SARS-CoV-2 biology, epidemiology, diagnosis, and potential management options. *Current Microbiology* 78(4), 1099–1114. https://doi.org/10.1007/s00284-021-02396-x

Bellini, S., Casadei, G., De Lorenzi, G., & Tamba, M. (2021). A review of risk factors of African swine fever incursion in pig farming within the European Union Scenario. *Pathogens* 10, 84. https://doi.org/10.3390/pathogens10010084.

Bhanuprakash, V., Indrani, B. K., Hosamani, M., & Singh, R. K. (2006). The current status of sheep pox disease. *Comparative Immunology, Microbiology and Infectious Diseases* 29, 27–60. https://doi.org/10.1016/j.cimid.2005.12.001

Bhatia, R., Narain, J. P., & Plianbangchang, S. (2012). Emerging infectious diseases in East and South-East Asia. In: R. Detels, S. G. Sullivan, & C. C. Tan (Eds.), *Public health in East and South-east Asia* (pp. 117–133). University of California Press.4

Blome, S., Gabriel, C., & Beer, M. (2013). Pathogenesis of African swine fever in domestic pigs and European wild boar. *Virus Research* 173, 122–130. https://doi.org/10.1016/j.virusres.2012.10.026.

Bora, B., Bora, D. P., Manu, M.,Barman, N. N., Dutta, L. J., Kumar, P. P., Poovathikkal, S., Suresh, K. P. & Nimmanapalli, K. (2020). Assessment of risk factors of African Swine Fever in India: Perspectives on future outbreaks and control strategies. *Pathogens*, 9, 1044. https://doi.org/10.3390/pathogens9121044

Bruno, L., Nappo, M. A., Ferrari, L., Di Lecce, R., Guarnieri, C., Cantoni, A. M., & Corradi, A. (2022). Nipah virus disease: Epidemiological, clinical, diagnostic and legislative aspects of this unpredictable emerging zoonosis. *Animals: An Open Access Journal from MDPI* 13(1), 159. https://doi.org/10.3390/ani13010159.

Cascella, M., Rajnik, M., Aleem, A., Dulebohn, S. C., & Di Napoli, R. (2023). Features, evaluation, and treatment of coronavirus (COVID-19). In *StatPearls*. StatPearls Publishing.

Centers for Disease Control and Prevention (CDC). (2022a). Nipah virus (NiV). Avaialble online: https://www.cdc.gov/vhf/nipah/index.html (accessed on 9 April 2023).

Centers for Disease Control and Prevention (CDC). (2022b). Mpox outbreak global map. Available online: https://www.cdc.gov/poxvirus/mpox/response/2022/world-map.html (accessed on 14 April 2023).

Chadha, M. S., Comer, J. A., Lowe, L., Rota, P. A., Rollin, P. E., Bellini, W. J., Ksiazek, T. G., & Mishra, A. (2006). Nipah virus-associated encephalitis outbreak, Siliguri, India. *Emerging Infectious Diseases* 12(2), 235–240. https://doi.org/10.3201/eid1202.051247

Chen, K.-T. (2022). Emerging infectious diseases and one health: Implication for public health. *International Journal of Environmental Research and Public Health* 19, 9081. https://doi.org/10.3390/ ijerph19159081

Ching, P. K., de los Reyes, V. C., Sucaldito, M. N., Tayag, E., Columna-Vingno, A. B., Malbas, F. F., Jr, Bolo, G. C., Jr, Sejvar, J. J., Eagles, D., Playford, G., Dueger, E., Kaku, Y., Morikawa, S., Kuroda, M., Marsh, G. A., McCullough, S., & Foxwell, A. R. (2015). Outbreak of henipavirus infection, Philippines, 2014. *Emerging Infectious Diseases* 21(2), 328–331. https://doi.org/10.3201/eid2102.141433

Chua, K. B. (2003). Nipah virus outbreak in Malaysia. *Journal of Clinical Virology: The Official Publication of the Pan American Society for Clinical Virology* 26(3), 265–275. https://doi.org/10.1016/s1386 -6532(02)00268-8

Chua, K. B., Bellini, W. J., Rota, P. A., Harcourt, B. H., Tamin, A., Lam, S. K., Ksiazek, T. G., Rollin, P. E., Zaki, S. R., Shieh, W., Goldsmith, C. S., Gubler, D. J., Roehrig, J. T., Eaton, B., Gould, A. R., Olson, J., Field, H., Daniels, P., Ling, A. E., Peters, C. J., ... Mahy, B. W. (2000). Nipah virus: a recently emergent deadly paramyxovirus. Science (New York, N.Y.), 288(5470), 1432–1435. https://doi.org/10.1126/science .288.5470.1432

Clemmons, E. A., Alfson, K. J., Dutton, J. W., III (2021). Transboundary animal diseases, an overview of 17 diseases with potential for global spread and serious consequences. *Animals* 11, 2039. https://doi.org/10 .3390/ ani11072039.

Cliff, A., Haggett, P., & Smallman-Raynor, M. (2000). *Island Epidemics Ch. 6* (pp. 165–236). Oxford University Press.

Coetzer, J. A. W. (2004). Lumpy skin disease. In Coetzer, J. A.W. Tustin, R. C (eds) *Infectious diseases of livestock*. Oxford University Press, Cape Town, 2nd edn., 2, 1268–1276.

Constable, P. D., Hinchcliff, K. W., Done, S. H., & Grundberg, W. (2017). *Veterinary medicine: A textbook of the diseases of cattle, horses, sheep, pigs, and goats* (11th ed., p. 1591). Elsevier.

Costard, S., Mur, L., Lubroth, J., Sanchez-Vizcaino, J. M., & Pfeiffer, D. U. (2013). Epidemiology of African swine fever virus. *Virus Research* 173, 191–197.

Daniels, P., Ksiazek, T., & Eaton, B. T. (2001). Laboratory diagnosis of Nipah and Hendra virus infections. *Microbes and Infection*, 3(4), 289–295. https://doi.org/10.1016/s1286-4579(01)01382-x

Daszak, P., Cunningham, A. A., Hyatt, A. D. (2000). Emerging infectious diseases of wildlife—threats to biodiversity and human health. *Science* 287, 443–449. https://doi.org/10.1111/tbed.12180.

Devnath, P., Wajed, S., Chandra Das, R., Kar, S., Islam, I., & Masud, H. M. A. A. (2022). The pathogenesis of Nipah virus: A review. *Microbial Pathogenesis* 170, 105693. https://doi.org/10.1016/j.micpath.2022 .105693

Dhama, K., Khan, S., Tiwari, R., Sircar, S., Bhat, S., Malik, Y. S., Singh, K. P., Chaicumpa, W., Bonilla-Aldana, D. K., & Rodriguez-Morales, A. J. (2020). Coronavirus disease 2019-COVID-19. *Clinical Microbiology Reviews* 33(4), e00028–20. https://doi.org/10.1128/CMR.00028-20

EFSA Panel on Animal Health and Welfare (AHAW), Nielsen, S. S., Alvarez, J., Bicout, D., Calistri, P., Depner, K., Drewe, J. A., Garin-Bastuji, B., Gonzales Rojas, J. L., Michel, V., et al. (2019). Risk assessment of African Swine fever in the South-eastern Countries of Europe. *EFSA Journal* 17.

El Eid, R., Allaw, F., Haddad, S. F., & Kanj, S. S. (2022). Human monkeypox: A review of the literature. *PLoS Pathogens* 18(9), e1010768. https://doi.org/10.1371/journal.ppat.1010768

Faltin, B., Zengerle, R., von Stetten, F. (2013). Current methods for fluorescence-based universal sequence-dependent detection of nucleic acids in homogenous assays and clinical applications. *Clinical Chemistry* 59, 1567–1582.

Fleming, D. T., et al. (1997). Herpes simplex virus type 2 in the United States, 1976 to 1994. *The New England Journal of Medicine* 337, 1105–1111. https://doi.org/10.1056/ NEJM199710163371601.

Food and Agriculture Organization of the United Nations. Transboundary Animal Diseases. (2021). Available online: www.fao.org/emergencies/emergency-types/transboundary-animal-diseases/en/ (accessed on 16 May 2021).

Givens, M. D. (2018). Review: Risks of disease transmission through semen in cattle. *Animal* 12(S1), s165–s171.

Guo, Z. D., Wang, Z. Y., Zhang, S. F., Li, X., Li, L., Li, C., Cui, Y., Fu, R. B., Dong, Y. Z., Chi, X. Y., Zhang, M. Y., Liu, K., Cao, C., Liu, B., Zhang, K., Gao, Y. W., Lu, B., & Chen, W. (2020). Aerosol and surface distribution of severe acute respiratory syndrome coronavirus 2 in hospital wards, Wuhan, China, 2020. *Emerging Infectious Diseases*, 26(7), 1583–1591. https://doi.org/10.3201/eid2607.200885.

Han, Q., Lin, Q., Ni, Z., & You, L. (2020). Uncertainties about the transmission routes of 2019 novel coronavirus. *Influenza and Other Respiratory Viruses* 14(4), 470–471. https://doi.org/10.1111/irv.12735

Harapan, H., Ophinni, Y., Megawati, D., Frediansyah, A., Mamada, S. S., Salampe, M., Bin Emran, T., Winardi, W., Fathima, R., Sirinam, S., Sittikul, P., Stoian, A. M., Nainu, F., & Sallam, M. (2022). Monkeypox: A comprehensive review. *Viruses* 14(10), 2155. https://doi.org/10.3390/v14102155

Hildebrandt, N., Spillmann, C. M., Algar, W. R., Pons, T., Stewart, M. H., Oh, E., Susumu, K., Diaz, S. A., Delehanty, J. B., & Medintz, I. L. (2017). Energy transfer with semiconductor quantum dot bio-conjugates: A versatile platform for biosensing, energy harvesting, and other developing applications. *Chemical Reviews* 117, 536–711.

Hsu, V. P., Hossain, M. J., Parashar, U. D., Ali, M. M., Ksiazek, T. G., Kuzmin, I., Niezgoda, M., Rupprecht, C., Bresee, J., & Breiman, R. F. (2004). Nipah virus encephalitis reemergence, Bangladesh. *Emerging Infectious Diseases* 10(12), 2082–2087. https://doi.org/10.3201/eid1012.040701

Hu, B., Guo, H., Zhou, P., & Shi, Z. L. (2021). Characteristics of SARS-CoV-2 and COVID-19. *Nature Reviews Microbiology* 19(3), 141–154. https://doi.org/10.1038/s41579-020-00459-7

Ince, Ö. B., Çakir, S., & Dereli, M. A. (2016). Risk analysis of lumpy skin disease in Turkey. *Indian Journal of Animal Research* 50(6), 1013–1017. 10.18805/ijar.9370

IOM (Institute of Medicine). (1992). *Emerging infections: Microbial threats to health in the United States.* National Academy Press.

Irons, P., Tuppurainen, E., & Venter, E. (2005). Excretion of lumpy skin disease virus in bull semen. *Theriogenology* 63, 1290–1297.

Jagadeesh, Bayry. (2013). Emerging viral diseases of livestock in the developing world. *Indian Journal of Virology* 24(3), 291–294. doi: 10.1007/s13337-013-0164-x.

Jin, Y. H., Cai, L., Cheng, Z. S., Cheng, H., Deng, T., Fan, Y. P., Fang, C., Huang, D., Huang, L. Q., Huang, Q., Han, Y., Hu, B., Hu, F., Li, B. H., Li, Y. R., Liang, K., Lin, L. K., Luo, L. S., Ma, J., Ma, L. L., … , for the Zhongnan Hospital of Wuhan University Novel Coronavirus Management and Research Team, Evidence-Based Medicine Chapter of China International Exchange and Promotive Association for Medical and Health Care (CPAM) (2020). A rapid advice guideline for the diagnosis and treatment of 2019 novel coronavirus (2019-nCoV) infected pneumonia (standard version). *Military Medical Research* 7(1), 4. https://doi.org/10.1186/s40779-020-0233-6

King, A. M., Adams, M. J., Carstens, E. B., & Lefkowitz, E. J. (2012). *Virus taxonomy. Classification and nomenclature of viruses.* Ninth Report of the International Committee on Taxonomy of Viruses, pp. 289–307

King, D. A., Peckham, C., Waage, J. K., Brownlie, J. & Woolhouse, M. E. J. (2006). Infectious diseases: Preparing for the future. *Science* 313, 1392–1393.

Ksiazek, T. G., Rota, P. A., & Rollin, P. E. (2011). A review of Nipah and Hendra viruses with an historical aside. *Virus Research* 162(1–2), 173–183. https://doi.org/10.1016/j.virusres.2011.09.026

Kugelman, J. R., Johnston, S. C., Mulembakani, P. M., Kisalu, N., Lee, M. S., Koroleva, G., McCarthy, S. E., Gestole, M. C., Wolfe, N. D., Fair, J. N., Schneider, B. S., Wright, L. L., Huggins, J., Whitehouse, C. A., Wemakoy, E. O., Muyembe-Tamfum, J. J., Hensley, L. E., Palacios, G. F., & Rimoin, A. W. (2014). Genomic variability of monkeypox virus among humans, Democratic Republic of the Congo. *Emerging Infectious Diseases* 20(2), 232–239. https://doi.org/10.3201/eid2002.130118

Kulkarni, D. D., Tosh, C., Venkatesh, G., et al. (2013). Nipah virus infection: current scenario. *Indian Journal of Virology* 24, 398–408. https://doi.org/10.1007/s13337-013-0171-y

Kumar, N., Chander, Y., Kumar, R., Khandelwal, N., Riyesh, T., Chaudhary, K., et al. (2021). Isolation and characterization of lumpy skin disease virus from cattle in India. *PLoS One* 16(1), e0241022. https://doi.org/10.1371/journal.pone.0241022

Laing, G., Vigilato, M. A. N., Cleaveland, S., Thumbi, S. M., Blumberg, L., Salahuddin, N., Abdela-Ridder, B., & Harrison, W. (2021). One Health for neglected tropical diseases. *Transactions of the Royal Society of Tropical Medicine and Hygiene* 115, 182–184.L

Lederberg, J., Shope, R. E., & Oaks, S. C. Jr (Eds.) (1992). *Emerging infections: microbial threats to health in the United States.* National Academy of Sciences.

Lei, H., Ye, F., Liu, X., Huang, Z., Ling, S., Jiang, Z., Cheng, J., Huang, X., Wu, Q., Wu, S., Xie, Y., Xiao, C., Ye, D., Yang, Z., Li, Y., Leung, N. H. L., Cowling, B. J., He, J., Wong, S. S., & Zanin, M. (2020). SARS-CoV-2 environmental contamination associated with persistently infected COVID-19 patients. *Influenza and Other Respiratory Viruses* 14(6), 688–699. https://doi.org/10.1111/irv.12783

Long, Q. X., Liu, B. Z., Deng, H. J., et al. (2020). Antibody responses to SARS-CoV-2 in patients with COVID-19. *Nature Medicine* 26, 845–848. https://doi.org/10.1038/s41591-020-0897-1

Lum, F. M., Torres-Ruesta, A., Tay, M. Z., Lin, R. T. P., Lye, D. C., Rénia, L., & Ng, L. F. P. (2022). Monkeypox: disease epidemiology, host immunity and clinical interventions. *Nature Reviews Immunology* 22(10), 597–613. https://doi.org/10.1038/s41577-022-00775-4

Ma, L., Chen, Z., Guan, W., Chen, Q., & Liu, D. (2019). Rapid and specific detection of all known Nipah virus Strains' sequences with reverse transcription-loop-mediated isothermal amplification. *Frontiers in Microbiology* 10, 418. https://doi.org/10.3389/fmicb.2019.00418

Magnus, P. V., Andersen, E. K., Petersen, K. B., & Birch-Andersen, A. (1959). A pox-like disease in cynomolgus monkeys. *Pathologica et Microbiologica Scandinavica* 46, 156–176. doi: 10.1111/j.1699-0463.1959. tb00328.x.

Malik, Y. S., & Dhama, K. (2015). Zika virus—an imminent risk to the world. *Journal of ImmunolImmunopathology* 17(2), 57–59. https://doi.org/10.5958/0973-9149.2015.00019.2

Maurer, F. D. (1962). Equine piroplasmosis – Another emerging disease. *Journal of the American Veterinary Medical Association* 141, 699–702.

Mbala, P. K., Huggins, J. W., Riu-Rovira, T., Ahuka, S. M., Mulembakani, P., Rimoin, A. W., Martin, J. W., & Muyembe, J. T. (2017). Maternal and fetal outcomes among pregnant women with human monkeypox infection in the democratic republic of Congo. *The Journal of Infectious Diseases* 216(7), 824–828. https://doi.org/10.1093/infdis/jix260

McCollum, A. M., & Damon, I. K. (2014). Human monkeypox. *Clinical Infectious Diseases: An Official Publication of the Infectious Diseases Society of America* 58(2), 260–267. https://doi.org/10.1093/cid /cit703

Menasherow, S., Rubinstein-Giuni, M., Kovtunenko, A., Eyngor, Y., Fridgut, O., Rotenberg, D., Khinich, Y., & Stram, Y. (2014). Development of an assay to differentiate between virulent and vaccine strains of lumpy skin disease virus (LSDV). *Journal of Virological Methods* 199, 95–101.

Molla, W., de Jong, M. C. M., & Frankena, K. (2017, November 6). Temporal and spatial distribution of lumpy skin disease outbreaks in Ethiopia in the period 2000 to 2015. *BMC Veterinary Research* 13(1), 310. doi: 10.1186/s12917-017-1247-5.

Montgomery, R. E. (1921). On a form of swine fever occurring in British East Africa (Kenya Colony). *Journal of Comparative Pathology and Therapeutics* 34, 159–191.

Morens, D. M. (1998). Acute haemorrhagic conjunctivitis: dealing with a newly emerging disease. *Pacific Health Dialog* 5, 147–153.

Morens, D. M., Folkers, G. K. & Fauci, A. S. (2004). The challenge of emerging and re-emerging infectious diseases. *Nature* 430, 242–249.

Moss, B. (2007). Poxviridae: The viruses and their replication. In D. M. Knipe & P. M. Howley (Eds.), *Fields virology* (pp. 2905–2946). Lippincott Williams & Wilkins.

Mulatu, E., & Feyisa, A. (2018). Review: Lumpy skin disease. *Journal of Veterinary Science and Technology* 9(535), 1–8. https://doi.org/10.4172/2157-7579.1000535

Namazi, F., & Tafti, A. K. (2021). Lumpy skin disease, an emerging transboundary viral disease: A review. *Veterinary Medicine and Science* 7, 888–896.

OIE. (2013). *World organization for animal health. Lumpy skin disease.* Technical Disease Card.

OIE. (2019). Chapter 3.8.1 African swine fever (infection with African swine fever virus). In *Manual of diagnostic tests and vaccines for terrestrial animals.* Health, W.O.f.A., ed. Office international des épizooties.

OIE. (2020). OIE-Listed Diseases, Infections and Infestations in Force in 2020. Available online: https://www .oie.int/en/animal-health-in-the-world/oie-listed-diseases-2020/13 August 2020.

OIE Terrestrial Manual. (2018). chapter 3.4.12, Lumpy skin disease (NB: Version adopted in May 2017).

Okello, J.J. (2011): Use of information and communication tools and services by rural grain traders: The case of Kenyan maize traders. In: *International Journal of ICT Research and Development in Africa* 2(2): 39–53.

Ong, S. W. X., Tan, Y. K., Chia, P. Y., Lee, T. H., Ng, O. T., Wong, M. S. Y., & Marimuthu, K. (2020). Air, surface environmental, and personal protective equipment contamination by severe acute respiratory syndrome coronavirus 2 (SARS-CoV-2) from a symptomatic patient. *JAMA* 323(16), 1610–1612. https:// doi.org/10.1001/jama.2020.3227.

Overgaauw, P. A. M., Vinke, C. M., Hagen, M. A. E. V., & Lipman, L. J. A. (2020). A one health perspective on the human-companion animal relationship with emphasis on zoonotic aspects. *International Journal of Environmental Research and Public Health* 7, 3789. [CrossRef]

Park, S. E. (2020). Epidemiology, virology, and clinical features of severe acute respiratory syndrome–coronavirus-2 (SARS-CoV-2; Coronavirus Disease-19). *Clinical and Experimental Pediatrics* 63(4), 119–124. https://doi.org/10.3345/cep.2020.00493

Patz, J. A., Campbell-Lendrum, D., Holloway, T., & Foley, J. A. (2005). Impact of regional climate change on human health. *Nature* 438, 310–317.

Peiris, J. S. M., et al. (2003). The severe acute respiratory syndrome. *New England Journal of Medicine* 349, 2431–2441.

Penrith, M. L., & Vosloo, W. (2009). Review of African swine fever: Transmission, spread and control. *Journal of the South African Veterinary Association* 80, 58–62.

Petersen, E., Kantele, A., Koopmans, M., Asogun, D., Yinka-Ogunleye, A., Ihekweazu, C., & Zumla, A. (2019). Human monkeypox: Epidemiologic and clinical characteristics, diagnosis, and prevention. *Infectious Disease Clinics of North America* 33(4), 1027–1043. https://doi.org/10.1016/j.idc.2019.03.001

Plowright, W., Thomson, G. R., & Neser, J. A. (1994). African swine fever. In: J. A. W. Coetzer, G. R. Thomson, & R. C. Tustin (Eds.), *Infectious diseases of livestock with special reference to Southern Africa* (pp. 567–599). Oxford University Press.

Rajsri, K. S., & Rao, M. (2022). A review of monkeypox: The new global health emergency. *Venereology* 1, 199–211. https://doi.org/10.3390/venereology1020014

Rao, T. V., & Bandyopadhyay, S. K. (2000, December). A comprehensive review of goat pox and sheep pox and their diagnosis. Animal Health Research Reviews 1(2), 127–136. doi: 10.1017/s1466252300000116.

Reis, A. L., Netherton, C., & Dixon, L. K. (2017). Unraveling the armor of a killer: Evasion of host defenses by African swine fever virus. *Journal of Virology* 91, e02338-16.

Rouby, S., & Aboulsoud, E. (2016). Evidence of intrauterine transmission of lumpy skin disease virus. *Veterinary Journal*, 209, 193–195.

Saijo, M., Ami, Y., Suzaki, Y., Nagata, N., Iwata, N., Hasegawa, H., Ogata, M., Fukushi, S., Mizutani, T., Iizuka, I., Sakai, K., Sata, T., Kurata, T., Kurane, I., & Morikawa, S. (2008). Diagnosis and assessment of monkeypox virus (MPXV) infection by quantitative PCR assay: Differentiation of Congo Basin and West African MPXV strains. *Japanese Journal of Infectious Diseases* 61(2), 140–142.

Sanchez, E. G., Quintas, A., Perez-Nunez, D., Nogal, M., Barroso, S., Carrascosa, A. L., & Revilla, Y. (2012). African swine fever virus uses macropinocytosis to enter host cells. *PLoSPathog* 8, e1002754.

Sánchez-Vizcaíno, J. M., Mur, L., Gomez-Villamandos, J. C., & Carrasco, L. (2015). An update on the epidemiology and pathology of African swine fever. *Journal of Comparative Pathology* 152(1), 9–21. https://doi.org/10. 1016/j.jcpa.2014.09.00

Sharma, V., Kaushik, S., Kumar, R., Yadav, J. P., & Kaushik, S. (2019). Emerging trends of Nipah virus: A review. *Reviews in Medical Virology* 29(1), e2010. https://doi.org/10.1002/rmv.2010

Singh, R. K., Dhama, K., Chakraborty, S., Tiwari, R., Natesan, S., Khandia, R., Munjal, A., Vora, K. S., Latheef, S. K., Karthik, K., Singh Malik, Y., Singh, R., Chaicumpa, W., & Mourya, D. T. (2019). Nipah virus: Epidemiology, pathology, immunobiology and advances in diagnosis, vaccine designing and control strategies–a comprehensive review. *The Veterinary Quarterly* 39(1), 26–55. https://doi.org/10.1080/01652176.2019.1580827

Singhal, T., Kabra, S. K., & Lodha, R. (2022). Monkeypox: A review. *Indian Journal of Pediatrics* 89(10), 955–960. https://doi.org/10.1007/s12098-022-04348-0

Skowron, K., Bauza-Kaszewska, J., Grudlewska-Buda, K., Wiktorczyk-Kapischke, N., Zacharski, M., Bernaciak, Z., & Gospodarek-Komkowska, E. (2022). Nipah virus-another threat from the world of zoonotic viruses. *Frontiers in Microbiology* 12, 811157. https://doi.org/10.3389/fmicb.2021.811157

Smolinski, M. S., Hamburg, M. A., & Lederberg, J. (2003). *Microbial threats to health: Emergence, detection, and response.* National Academies Press.

Soman Pillai, V., Krishna, G., & ValiyaVeettil, M. (2020). Nipah virus: Past outbreaks and future containment. *Viruses* 12(4), 465. https://doi.org/10.3390/v12040465

Stobiecka, M., & Chalupa, A. (2015). Biosensors based on molecular beacons. *Chemical Papers* 69, 62–76.

Sudhakar, S. B., Mishra, N., Kalaiyarasu, S., Jhade, S. K., Hemadri, D., Sood, R., Bal, G. C., Nayak, M. K., Pradhan, S. K., & Singh, V. P. (2020). Lumpy skin disease (LSD) outbreaks in cattle in Odisha state, India in August 2019: Epidemiological features and molecular studies. *Transboundary and Emerging Diseases* 67(6), 2408–2422. https://doi.org/10.1111/tbed.13579

Tan, C. C. S., Lam, S. D., Richard, D., Owen, C. J., Berchtold, D., Orengo, C., Nair, M. S., Kuchipudi, S. V., Kapur, V., van Dorp, L., & Balloux, F. (2022). Transmission of SARS-CoV-2 from humans to animals and potential host adaptation. *Nature Communications* 13(1), 2988. https://doi.org/10.1038/s41467-022-30698-6

Tavakoli, A., Niya, M. H. K., Keshavarz, M., Ghaffari, H., Asoodeh, A., Monavari, S. H., & Keyvani, H. (2017). Current diagnostic methods for HIV. *Future Virology* 12, 141–155

Taylor, L. H., Latham, S. M., & Woolhouse, M. E. (2001). Risk factors for human disease emergence. *Philosophical Transactions of the Royal Society of London. Series B, Biological Sciences* 356, 983–989.

Thomas, L. (2002). Lumpy-skin disease, a disease of socioeconomic importance. *Journal of Medical Virology* 76, 6054–6061.

Titanji, B. K., Tegomoh, B., Nematollahi, S., Konomos, M., & Kulkarni, P. A. (2022). Monkeypox: A contemporary review for healthcare professionals. *Open Forum Infectious Diseases* 9(7), ofac310. https://doi.org/10.1093/ofid/ofac310

Tulman, C. L., Afonso, Z. L. U., Zsak, L., Kutish, G. F., & Rock, D. L. (2002). The genomes of sheeppox and goatpox viruses. *Journal of Virology* 76(12), 6054–6061. https://doi.org/10.1128/JVI.76.12.6054-6061.2002.

Tuppurainen, E. S., Venter, E. H., Shisler, J. L., Gari, G., Mekonnen, G. A., Juleff, N., Lyons, N. A., De Clercq, K., Upton, C., Bowden, T. R., Babiuk, S., & Babiuk, L. A. (2017). Review: Capripoxvirus diseases: Current status and opportunities for control. *Transboundary and Emerging Diseases* 64(3), 729–745. https://doi.org/10.1111/tbed.12444

Vemula, S. V., Zhao, J. Q., Liu, J. K., Wang, X., Biswas, S., Hewlett, I. (2016). Current approaches for diagnosis of influenza virus infections in humans. *Viruses* 8, 96.

Wang, D., Hu, B., Hu, C., Zhu, F., Liu, X., Zhang, J., Wang, B., Xiang, H., Cheng, Z., Xiong, Y., Zhao, Y., Li, Y., Wang, X., & Peng, Z. (2020a). Clinical characteristics of 138 hospitalized patients with 2019 novel coronavirus-infected pneumonia in Wuhan, China. *JAMA* 323(11), 1061–1069. https://doi.org/10.1001/jama.2020.1585

Wang, L., Shang, J., Weng, S., Aliyari, S. R., Ji, C., Cheng, G., & Wu, A. (2022). Genomic annotation and molecular evolution of monkeypox virus outbreak in 2022. *Journal of Medical Virology* 95(1), e28036. https://doi.org/10.1002/jmv.28036

Weiss, R. A., & McMichael, A. J. (2004). Social and environmental risk factors in the emergence of infectious diseases. *Nature Medicine* 10, S70–S76.

World Health Organization (WHO). (2022). Laboratory testing for the monkeypox virus: Interim guidance. Available online: https://www.who.int/publications/i/item/WHO-MPX-laboratory-2022.1 (accessed on 14 April 2023)

World Health Organization, Regional Office for South East Asia Region. (2005). *Combating emerging infectious diseases in the South-East Asia Region.* World Health Organization, WHO SEARO.

World Organization for Animal Health. (2021). *Animal diseases.* World Organization for Animal Health. Available online: www.oie.int/en/what-we-do/animal-health-and-welfare/animal-diseases/ (accessed on 16 May 2021).

Wu, T., Kang, S., Peng, W., Zuo, C., Zhu, Y., Pan, L., Fu, K., You, Y., Yang, X., Luo, X., Jiang, L., & Deng, M. (2021). Original hosts, clinical features, transmission routes, and vaccine development for coronavirus disease (COVID-19). *Frontiers in Medicine* 8, 702066. https://doi.org/10.3389/fmed.2021.702066

Xie, C., Zhao, H., Li, K., Zhang, Z., Lu, X., Peng, H., Wang, D., Chen, J., Zhang, X., Wu, D., Gu, Y., Yuan, J., Zhang, L., & Lu, J. (2020). The evidence of indirect transmission of SARS-CoV-2 reported in Guangzhou, China. *BMC Public Health* 20(1), 1202. https://doi.org/10.1186/s12889-020-09296-y

Yang, Z., Reynolds, S. E., Martens, C. A., Bruno, D. P., Porcella, S. F., & Moss, B. (2011). Expression profiling of the intermediate and late stages of poxvirus replication. *Journal of Virology* 85, 9899–9908.

Yu, J., Chai, P., Ge, S., & Fan, X. (2020). Recent understandings toward coronavirus disease 2019 (COVID-19): From bench to bedside. *Frontiers in Cell and Developmental Biology* 8, 476. https://doi.org/10.3389/fcell.2020.00476

Zhai, P., Ding, Y., Wu, X., Long, J., Zhong, Y., & Li, Y. (2020). The epidemiology, diagnosis and treatment of COVID-19. *International Journal of Antimicrobial Agents* 55(5), 105955. https://doi.org/10.1016/j.ijantimicag.2020.105955

Zhao, J., Cui, W., & Tian, B. P. (2020). The potential intermediate hosts for SARS-CoV-2. *Frontiers in Microbiology* 11, 580137. https://doi.org/10.3389/fmicb.2020.580137

Zheng, M., Liu, Q., Jin, N. Y., Guo, J. G., Huang, X., Li, H. M., Zhu, W., & Xiong, Y. (2007). A duplex PCR assay for simultaneous detection and differentiation of Capripoxvirus and Orf virus. *Molecular and Cellular Probes* 21, 276–281. https://doi.org/10.1016/j.mcp.2007.01.005

20 Immunity to Virus Infection

Madhuri Subbiah and Haajira Beevi Habeeb Rahuman
National Institute of Animal Biotechnology, Hyderabad, India

Viruses utilize their hosts for their propagation and to spread to other hosts; hence, viruses are defined as obligate intracellular parasites. The host responds to the viral infection with a complex defense mechanism that includes innate and adaptive immune responses for clearing the virus infection and preparing the host to protect itself from future attacks by the virus. In some cases, despite the immune responses mounted by the host, the viruses become lethal, causing the death of the host. This is observed when viruses jump hosts, alter their antigenicity, or when the host is immunocompromised. Many research studies on the host response to virus infection have helped us understand concepts such as immunological tolerance, antigen processing and presentation, and to some extent, even cancer and autoimmunity.

HOST RESPONSES TO VIRAL INFECTION

Viruses enter the host through skin, blood, respiratory, oral, nervous, or genital tracts. Mechanical or anatomical barriers such as the skin, pH of the mucous membranes, and the normal microflora living in the skin and mucous membranes, act first during viral infections. It is becoming increasingly clear that the commensal microflora have a subtle yet important role in antiviral immune response, which can influence the outcome of virus infection. Additionally, physiological barriers such as body temperature, gastric acidity of the stomach, lysozyme in tears and mucus, interferon, and collectins, which are the surfactant proteins in serum and lung secretions, also act to defend against the entering viruses (Figure 20.1a).

Viruses enter and replicate in certain tissues depending on the availability of suitable receptors on those cells. Once inside the cells, the viruses utilize the host machinery for their own propagation. Some viruses induce cell death to release new infectious virions (e.g., poxviruses, polioviruses, and herpesviruses), while in other viruses, like HIV and flu viruses, the virions bud from the infected cells without causing cell death.

During their entry, replication, and subsequent progeny virion release, the viruses encounter innate immune defenses aimed at clearing the virus infection and also activating the adaptive immunity. The innate immunity is an immediate and early defense against the virus without any specificity; however, the adaptive immunity is a delayed response yet more specific in nature.

INNATE IMMUNITY

Innate defenses mediated by immune cells such as dendritic cells block the initial infection and eliminate virus-infected cells. This is achieved via sensors called pathogen recognition receptors of the Toll-like receptor (TLR) family or RNA helicase family (1). TLRs are either expressed on the cell surface or are endosomal membrane-bound on dendritic cells (DCs), macrophages, lymphocytes, and parenchymal cells (2). While some of the TLRs (TLR 7/8/9) are expressed constitutively at high levels by plasmacytoid DCs, other TLRs are inducible upon viral stimuli (single- and double-stranded RNA stimulate expression of TLR 3 and TLR 7/8, respectively, double-stranded DNA induces TLR 9 expression). RNA helicases, retinoic acid-inducible gene I (RIG-I), and melanoma differentiation-associated gene 5 (MDA-5), as well as cytosolic dsDNA sensor DAI

Immunity to Virus Infection

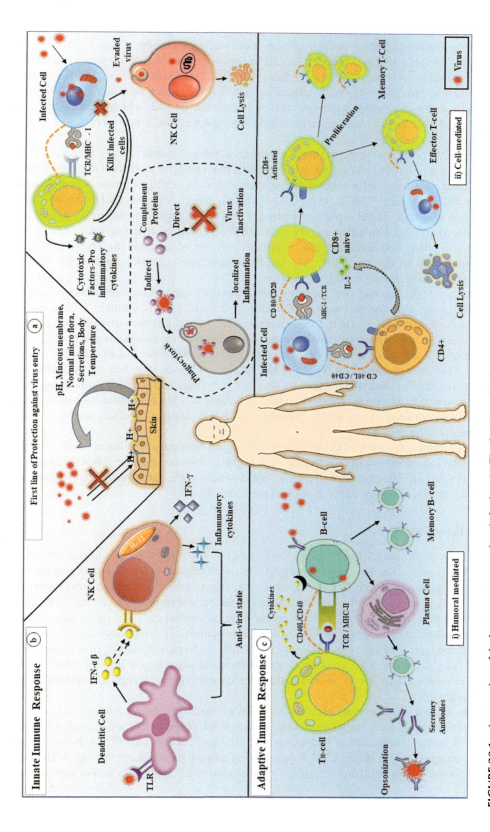

FIGURE 20.1 An overview of the host responses to virus infection. (a) The first line of defense during virus entry includes anatomical and physiological barriers that protect the host from virus entry. (b) Innate immune responses, and the initial fast but nonspecific response to virus infection. (c) The adaptive immune response, which is slow but specific and can develop memory for future encounters of the same virus.

(DNA-dependent activator of interferon-reulatory factors), can recognize viruses (3). Recognition of the viral presence through these cellular sensors leads to the generation of type I (α/β) interferons (IFNs), which bind to IFN receptors and result in the expression of IFN-stimulated genes and inflammatory cytokines (4). This sets the cells in an "antiviral" state wherein cell protein synthesis is shut off, which in turn prevents viral replication. Type I IFNs induce natural killer (NK) cells directly or indirectly by inducing cytokines such as interleukin (IL)-12, which induces NK cells to secrete IFN-γ. Other innate immune cells include macrophages, DCs, neutrophils, and T cells expressing γ δ T-cell receptors.

Apart from IFNs, other host proteins that play important roles in innate defense are natural antibodies, a natural source of circulating immunoglobulins that can bind to viruses (5), complement proteins that clear the infection by either direct virus inactivation or phagocytosis of complement-bound virions (6), cytokines (TNF-α, IFN-γ, IL-12, IL-6), and chemokines (MIP-1α), which are induced by virus infection. The complement pathways lead to phagocytosis and trigger a localized inflammatory response (7). The inflammatory response allows immune reactions (exhibited as redness, heat, pain, swelling, and functional loss) to initiate at the site of infection (8). The orchestration of innate immune cells to the site of infection and their subsequent retention for effective interactions is executed by these inflammatory chemokines.

The infected cells display the viral antigens through class I major histocompatibility complex proteins (MHC class I) on the cell surface. Circulating T cells, called cytotoxic T cells, recognize these virus-infected cells via their T-cell receptors (TCRs) and kill the virus-infected cells by releasing cytotoxic factors. However, some viruses evade the host by minimizing the MHC class I presentation on the cell surface; these cells are recognized by NK cells and are killed. The pro-inflammatory cytokines produced by NK cells kill infected cells and also help to associate with DCs. DCs are regarded as the cellular sentinels that link the innate and adaptive immune systems. This sentinel role is very important for the first line of defense at virus entry sites such as the skin, and gastrointestinal and respiratory tracts. Subsets of DCs with distinct functions are localized in different tissues (Figure 20.1b) (9).

The innate immune response blocks the virus infection and dissemination. Innate immunity may by itself clear the infection or help in activating the adaptive immune response. The role played by innate immunity has an influence on the type and effectiveness of the subsequent adaptive immune response, which comes into play within a week of virus infection.

ADAPTIVE IMMUNITY

Adaptive immune responses are activated upon interaction of the virus with the antigen-presenting cells or B-cell receptors. The two arms of adaptive immune responses are (i) humoral immunity mediated by antibodies produced by B cells, which are effective in blocking the virus entry into cells during the early stages of virus infection or during the release of progeny viruses, or by causing the death of infected cells, all of which is action directed against the virus in the extracellular space; and (ii) cell-mediated immunity (CML), driven by cytotoxic T lymphocytes (CTLs), which have a surveillance role and kill virus-infected cells. CML response is raised against intracellular pathogens. T cells (CD4 and CD8) mature in the thymus and get released into the bloodstream. During virus infection, CD8+ T cells are CTLs that identify cytosolic viral peptides on class I MHC molecules (Figure 20.1c).

B cells are produced in the bone marrow and then moves to the lymph nodes. Following an antigen encounter, the B cells become activated. Short-lived plasma cells are activated through upregulation of certain transcription factors such as Blimp-1, XBP-1, and IRF-4. Some B cells interact with antigen-specific helper CD4 T cells during antigen presentation through MHC class II molecules. Somatic hypermutation and affinity maturation result in the selection of high-affinity

Immunity to Virus Infection

and long-lived plasma cells and memory B cells (10). IgM is the primary antibody produced during an immune response. With progression, the activated plasma cells produce IgG antibodies, which are highly specific and thus important antibodies during vaccination. IgD is a receptor-bound antibody, while IgA is a secretory antibody in mucus membranes. IgE is found during allergic reactions and parasitic infections.

Binding of viruses by specific antibodies can prevent the virus from infecting a cell. This binding can effectively neutralize the virus, cause aggregation of viral particles, and trigger phagocytosis, by which the cell recognizes the antibody-bound viral particle and consumes the virus. The complement system induces phagocytosis of viruses that are bound by antibodies and also perforates the phospholipid bilayer of enveloped viruses.

IgA and IgM are actively transported across the mucosal epithelium and can neutralize viruses intracellularly. Mucosal secretory IgA (sIgA) and systemic IgG are effective against viruses. Thus, antibodies find their use as therapeutics to treat viral diseases. For measuring the efficacy of some vaccines, induction of neutralizing antibodies determines the degree of protection against disease (11).

The most antigenic viral proteins that induce antibodies are generally outer surface proteins, and viruses escape antibody binding by altering their viral antigen gene, which is commonly observed in HIV and influenza viruses.

T cells further play a role in humoral immunity by supporting the signals for B cell activation, maturation, and memory of the humoral response. T-cell immunity includes CD4 and CD8 T-cell subsets that recognize viral peptides bound to surface MHC proteins class II and class I, respectively. Unlike B cells, T cells require the antigen to be processed and presented by antigen-presenting cells either as intracellular antigens tagged along with MHC I proteins, which is generally noticed in viral infections, or by extracellular antigens tagged with MHC II proteins, noticed in bacterial infections. Once activated, the T cells clonally expand, producing effector T cells to clear the infection and memory T cells to fight the infections with the same antigen in the future.

The precursors of specific CD8 cytotoxic T lymphocytes (CTLs) upon recognition of viral antigens on MHC class I get activated through co-stimulatory molecules (CD28 and 4-1BB) and release inflammatory cytokines (IFN-α/β and IL-12), which enable them to proliferate and differentiate into effectors (12). Killing of virus-infected cells by CTLs occurs by apoptosis and involves perforins and cytotoxic granules, and through Fas ligand attachment with Fas on infected cells. In the case of non-rejuvenating cells getting infected with a virus, for example, in HSV infection of nerve cells, purging of the virus from infected cells without killing happens through the release of cytokines such as IFN-γ, TNF-α, lymphotoxin-α, and RANTES by CD8 T cells. The four phases of CTL antiviral responses are as follows: (1) induction phase, wherein the CD8+ precursor T cells proliferate and differentiate into effector cells that kill the virus-infected cells; (2) activation-induced cell-death (AICD) phase, wherein the activated CTLs undergo apoptosis once the virus is cleared; (3) silencing phase, wherein there is no activation and apoptosis continues in virus-specific CTLs; and (4) memory phase wherein some of the virus-specific activated CTLs retain their viability and undergo dormancy as memory cells with the ability to proliferate upon reinfection with the same virus and to kill the virus-infected cells.

CD4 T cells' role as helper cells is important in certain virus infections such as HSV, HIV, and influenza virus by stimulating both antiviral antibodies and CD8 T-cell responses and generating cytokines for the virus clearance from infected cells. The effector pool of CD4 T cells is smaller than that observed with CD8 T cells. CD4 T cells aid in the production of long-lived memory CD8 T cells. CD4 subset effector cells, T-helper 1 (Th1) cells, produce inflammatory cytokines IFN-γ, LTα, TNF-α, and IL-2, while Th2 effectors release IL-4, IL-5, and two anti-inflammatory cytokines, IL-10 and transforming growth factor-β (TGF-β), to counter the Th1 cells. IL-17-producing subsets of CD4 T cells are associated with immune pathogenesis.

IMMUNOLOGICAL MEMORY

It is very important for the host to remember the previous infection and quickly fight if attacked by the same virus another time. This is immunological memory, which is driven by adaptive immunity. Immunological memory is marked by the availability of specific lymphocytes and antibody-producing plasma cells in bone marrow, which quickly increase in numbers upon infection with a previously encountered specific virus enabling the body to fight back the virus without severe illness. This forms the basis for vaccination (13). Immunological memory is long-lived for certain viruses such as measles and smallpox.

The clonal expansion and clonal differentiation form part of immunological memory in B cells. The primary humoral response is predominantly characterized by closely related antibodies showing minimal somatic hypermutations; however, numerous somatic hypermutations are seen in antibodies along with antibodies containing VH and VL gene segments during the secondary immune response, which were not detected before. The stimulation of B cells while encountering the same virus for the second time happens more efficiently for the higher-affinity memory B cells.

Cytokines IL-7 and IL-15 control the homeostatic division of memory T cells. Repeated stimulation of memory cells, either due to repeated infections with the same virus or vaccine boosters, is known to increase the antigen-specific memory T-cell pool (14). Memory T cells are of two tissue-specific subsets: effector memory (TEM) and central memory (TCM) subsets. The CD62LloCCR7lo TEM subsets are primarily in nonlymphoid tissues and spleen, while the CD62LhiCCR7hi TCM subsets are seen in the lymph nodes and spleen. Immunological memory may contribute to autoimmunity.

INBORN ERRORS OF IMMUNITY (IEIS)

IEIs refer to either developmental or functional disorders of immune cells. Patients with IEIs suffer from severe viral infections and are prone to life-threatening illnesses, including malignancies, upon virus infections. Studies on IEIs have infact revealed many antiviral host mechanisms and provided a deeper understanding of host–virus interactions and viral pathogenesis. This knowledge is helpful in treating IEI patients using specific therapies, for example, SCID (severe combined immunodeficiency), wherein the T lymphocytes are mutated and there is an absence of B and NK cells. Studies of SCID patients helped to find the roles of T-cell immunity in virus infections (15). In COVID-19 patients, those with severe disease and with low serum type IFN had defective IFN type I TLR3 and the IRF-7 dependent pathway. Type I IFN therapy in these patients was found beneficial (16).

IMMUNE EVASION

Many viruses inhibit IFN signaling and production: the NS3-4A protease of the hepatitis C virus inhibits interferon production by blocking IRF3 phosphorylation; the V proteins of paramyxoviruses block downstream signaling pathways leading to IFN inhibition.

Viruses can escape immune cells by modifying their surface antigens. Such antigenic modifications via point mutations (antigenic drift) or by recombination of their genes (antigenic shift) are commonly seen in influenza viruses.

Another immune evasion strategy is latency and persistence. In latent infections, the virus stays longer in the cell without replicating, hence not attracting the attention of immune cells (herpesviruses). During favorable conditions such as immunosuppression, the virus may reactivate and spread, leading to severe disease. Persistent infection means a chronic infection that causes host tissue damage and/or hyperinflammations (macrophage activation syndrome, polyclonal lymphoid proliferations).

Immunity to Virus Infection

Viruses are also known to express genes homologous to the receptors of some cytokines or modulate or inhibit cytokines and chemokines. Viruses may delay or block the CTL-induced apoptosis of infected cells, inhibit antigen processing, and downregulate MHC molecules on the surface of infected cells. Antigenic hypervariability is noticed in some viral infections to escape from antibody or T-cell recognition (influenza viruses).

IMMUNOPATHOLOGY

During viral infections, the host counteracts the virus through immune responses but with least damage to host tissues. However, in some cases, infections with HIV, hepatitis C virus (HCV), hepatitis B virus (HBV), and herpesviruses, immunopathology or damage to host tissues are noticed. This means developing effective vaccines against these viruses will be tricky (17). Lethal cytokine storms are reported in infections with influenza viruses and the recent coronavirus infections in many untreated cases. Viruses such as coxsackie B virus, human T lymphotropic virus, and Epstein–Barr virus are known to cause cancer in certain patients.

The extent of the tissue damage also depends on the host's genetics, physiological status, age, viral infection dose, route, and sometimes co-infections. The tissue damage could be contributed by both innate and adaptive immune arms of the host. Excessive activities of the pro-inflammatory cytokines by the innate immune system can lead to tissue damage. Persistent viral infection, viral evasion strategies that inhibit IFN production, or those interfering with antigen processing and presentation lead to host tissue damage. Further, adaptive immune effector cells, such as T cells, may directly damage the virus-infected cell or indirectly by releasing cytokines such as tumor necrosis factor. TH17 cells lead to inflammatory responses during HIV, HCV, and influenza virus infections via neutrophil recruitment and cause tissue damage. Antibody-mediated inflammatory reactions may also contribute to tissue damage. In West Nile infection, the inflammatory responses break down the blood–brain barrier, which leads to brain infection augmenting the lethality of the viral infection.

One may wonder how the host is going to counteract the tissue damage. Cytokines such as IL-10, TGFβ, and IL-17 have anti-inflammatory activity. Anti-inflammatory mediators include resolvins, galectins, and protectins, and certain cell subsets, such as T reg cells, inhibit other cells from mediating inflammatory events. These cells produce certain molecules such as programmed cell death protein 1 (PD1) and the IL-10 receptor during chronic infections that inhibit effector cells, termed the exhaustion phenotype. PD1-induced high levels of IL-10 production in monocytes has been shown to inhibit the functions of CD4+ T cells (18).

Heterologous virus immunity refers to an immune response mounted against a particular pathogen cross-reacting with another virus. Though the cross-reactivity may be of low avidity and barely protective, it could cause tissue damage. This explains the severe condition called dengue hemorrhagic fever observed in some individuals who are already immune to one strain of dengue virus and get infected with another strain. Similarly, heterologous immunity leading to the debilitating disease infectious mononucleosis is presented in a smaller fraction of young adults who have antibodies to the Epstein–Barr virus from an earlier infection. Antibody-dependent cellular cytotoxicity is noticed in dengue virus and HIV infections wherein the Fc receptor binding of antibody-bound viruses may favor further infection, causing more severe tissue damage (19).

A deeper knowledge of the molecular basis of immunopathology during viral infection will be helpful to devise control and prevention strategies either during or prior to viral infections specifically helpful in the event of epidemics.

AUTOIMMUNITY

Breakdown of self-tolerance due to aberrant immune response is implicated in autoimmune diseases. Several factors are hypothesized to lead to autoimmunity, for example molecular mimicry

Textbook of General Virology

wherein the virus shares epitopes with the host, so antibodies produced to target the virus will now react with host cells as well. The prolonged release of cytokines during viral infection, specifically in persistent infections are also implicated in inducing auto immunity (20, 21).

VIRUS INFECTION CAUSING IMMUNE SUPPRESSION

Infection with certain viruses is known to cause immunosuppression, which sometimes is transient. The possible mechanisms could be infection of the immune cells, defective co-stimulation, DC dysfunction, expression of suppressive factors (anti-inflammatory cytokines), or stimulation of suppressor cells, majorly executed through certain viral protein expressions or through skewed host defense mechanisms. For example, the kinetics of IFN is essential in determining whether IFN inhibits or augments T-cell proliferation (22). The immune dysfunction noticed in HIV-infected patients is caused by infection and rapid destruction of CD4 T cells, together with reduced thymic production leading to acquired immune deficiency syndrome (AIDS) (23). Immune suppression makes the host susceptible to other infections. It is speculated that the immune suppression may be causing the long COVID symptoms.

HOST IMMUNITY AND VACCINE DEVELOPMENT

With the knowledge of host–virus interactions, it becomes possible to develop effective vaccines against viral diseases. Edward Jenner's attempt with cowpox to stimulate smallpox immunity in 1798 is a best example . The goal of vaccination is to trigger innate immune responses that subsequently lead to specific adaptive immune responses against the virus through the production of effector cells and memory cells (24, 25).

REFERENCES

1. Kawai, Taro, and Shizuo Akira. "TLR signaling." *Cell Death & Differentiation* 13, no. 5 (2006): 816–825.
2. Kopp, Elizabeth, and Ruslan Medzhitov. "Recognition of microbial infection by Toll-like receptors." *Current Opinion in Immunology* 15, no. 4 (2003): 396–401.
3. Kawai, Taro, and Shizuo Akira. "Innate immune recognition of viral infection." *Nature Immunology* 7, no. 2 (2006): 131–137.
4. García-Sastre, Adolfo, and Christine A. Biron. "Type 1 interferons and the virus-host relationship: A lesson in detente." *Science* 312, no. 5775 (2006): 879–882.
5. Ochsenbein, Adrian F., Thomas Fehr, Claudia Lutz, Mark Suter, Frank Brombacher, Hans Hengartner, and Rolf M. Zinkernagel. "Control of early viral and bacterial distribution and disease by natural antibodies." *Science* 286, no. 5447 (1999): 2156–2159.
6. Ahmed, R., and C. A. Biron, eds. *Immunity to Viruses.* Philadelphia: Lippincott-Raven, 1998.
7. The Merck Manuals Online Medical Library. 2008. "Complement system." [Last cited on 2009 Nov 22]. http://www.merck.com/mmpe/sec13/ch163/ch163d.html
8. Goldsby, R. A., T. J. Kindt, B. A. Osborne, and J. Kuby. "Chapter 2: Cells and organs of the immune system." In *Immunology,* edited by Goldsby, R. A., T. J. Kindt, A. Osborne and J. Kuby., 5th ed. New York: W. H. Freeman and Company, 2003.
9. Biron, Christine A., Khuong B. Nguyen, Gary C. Pien, Leslie P. Cousens, and Thais P. Salazar-Mather. "Natural killer cells in antiviral defense: Function and regulation by innate cytokines." *Annual Review of Immunology* 17, no. 1 (1999): 189–220.
10. McHeyzer-Williams, Louise J., Laurent P. Malherbe, and Michael G. McHeyzer-Williams. "Checkpoints in memory B-cell evolution." *Immunological Reviews* 211, no. 1 (2006): 255–268.
11. Whitton, J. L., and M. B. A. Oldstone. "The immune response to viruses." In *Fields Virology,* edited by D. M. Knipe and P. M. Howley, 285–320, 4th ed. Philadelphia: Lippincott Williams and Wilkins, 2001.
12. Mescher, Matthew F., Julie M. Curtsinger, Pujya Agarwal, Kerry A. Casey, Michael Gerner, Christopher D. Hammerbeck, Flavia Popescu, and Zhengguo Xiao. "Signals required for programming effector and memory development by CD8+ T cells." *Immunological Reviews* 211, no. 1 (2006): 81–92.

13. Janeway, C. A. Jr., P. Travers, M. Walport, et al. *Immunobiology: The Immune System in Health and Disease.* 5th ed. New York: Garland Science, 2001.

14. Masopust, David, Sang-Jun Ha, Vaiva Vezys, and Rafi Ahmed. "Stimulation history dictates memory CD8 T cell phenotype: implications for prime-boost vaccination." *The Journal of Immunology* 177, no. 2 (2006): 831–839.

15. Cirillo, Emilia, Caterina Cancrini, Chiara Azzari, Silvana Martino, Baldassarre Martire, Andrea Pession, Alberto Tommasini et al. "Clinical, immunological, and molecular features of typical and atypical severe combined immunodeficiency: Report of the italian primary immunodeficiency network." *Frontiers in Immunology* 10 (2019): 1908.

16. Zhang, Qian, Paul Bastard, Zhiyong Liu, Jérémie Le Pen, Marcela Moncada-Velez, Jie Chen, Masato Ogishi et al. "Inborn errors of type I IFN immunity in patients with life-threatening COVID-19." *Science* 370, no. 6515 (2020): eabd4570.

17. Rouse, Barry T., and Sharvan Sehrawat. "Immunity and immunopathology to viruses: what decides the outcome?" *Nature Reviews Immunology* 10, no. 7 (2010): 514–526.

18. Said, Elias A., Franck P. Dupuy, Lydie Trautmann, Yuwei Zhang, Yu Shi, Mohamed El-Far, Brenna J. Hill et al. "Programmed death-1–induced interleukin-10 production by monocytes impairs CD4+ T cell activation during HIV infection." *Nature Medicine* 16, no. 4 (2010): 452–459.

19. Takada, Ayato, and Yoshihiro Kawaoka. "Antibody-dependent enhancement of viral infection: Molecular mechanisms and in vivo implications." *Reviews in Medical Virology* 13, no. 6 (2003): 387–398.

20. Getts, Daniel R., Emily ML Chastain, Rachael L. Terry, and Stephen D. Miller. "Virus infection, antiviral immunity, and autoimmunity." *Immunological Reviews* 255, no. 1 (2013): 197–209.

21. Mueller, Scott N., and Barry T. Rouse. "Immune responses to viruses." *Clinical Immunology* 15, (2008): 421.

22. Marshall, Heather D., Stina L. Urban, and Raymond M. Welsh. "Virus-induced transient immune suppression and the inhibition of T cell proliferation by type I interferon." *Journal of Virology* 85, no. 12 (2011): 5929–5939.

23. Dion, Marie-Lise, Jean-François Poulin, Rebeka Bordi, Myriam Sylvestre, Rachel Corsini, Nadia Kettaf, Ali Dalloul et al. "HIV infection rapidly induces and maintains a substantial suppression of thymocyte proliferation." *Immunity* 21, no. 6 (2004): 757–768.

24. Clem, Angela S. "Fundamentals of vaccine immunology." *Journal of Global Infectious Diseases* 3, no. 1 (2011): 73.

25. Braciale, Thomas J., and Young S. Hahn. "Immunity to viruses." *Immunological Reviews* 255, no. 1 (2013): 5.

21 Viral Vaccines

Vijay Singh Bohara and Sachin Kumar
Department of Biosciences and Bioengineering,
Indian Institute of Technology Guwahati, Guwahati, Assam, India

INTRODUCTION

Vaccinations are reliable means of protection from a variety of potentially fatal illnesses. The term "vaccine" is derived from the Latin word *vaccinus,* meaning "from cows." Vaccines are used to generate active immunity against infectious diseases or pathogens. In vaccines, antigens or a combination of antigens that do not cause disease are used to stimulate the body's immune system. Multiple successful vaccinations against several viral diseases have been developed through the study of vaccinology, reducing the severity of some severe viral diseases.

A BRIEF HISTORICAL PERSPECTIVE OF VIRAL VACCINES

The first vaccine was developed by Edward Jenner in 1796 against smallpox. He inoculated a boy with pus from the cowpox lesions of a milkmaid. Later, he inoculated the boy with human pus from smallpox lesions. He observed that post-inoculation, the boy did not show any smallpox symptoms and remained healthy (Saleh et al. 2021). In 1885, Louis Pasteur developed the first vaccine against rabies (Bordenave 2003). In 1937, Max Theiler, Hugh Smith, and Eugen Haagen designed the first vaccine against yellow fever (Frierson 2010). In 1946, the first vaccine against influenza virus infection was approved for public use. It was developed by Thomas Francis Jr. and Jonas Salk (Francis, Salk, and Quilligan 1947). Jonas Salk also developed the first effective vaccine against polio. The second polio vaccine was developed by Albert Sabin. It was the first live-attenuated vaccine that could be given orally as drops (Blume and Geesink 2000). The first hepatitis B virus vaccine was developed by microbiologist Irving Millman. It was a heat-killed form of the virus (Coffin 2022). In 2006, the first vaccine for human papillomavirus (HPV) was approved and played a key role in the elimination of cervical cancer (Ljubojevic 2006).

STRATEGIES FOR VIRAL VACCINE DEVELOPMENT

A successful vaccination must adhere to three criteria: it must be safe, it must prevent infection effectively, and the approach must be acceptable to the community at risk. Traditional vaccinations, often known as first-generation vaccines, use either a pure antigen or the whole organism as their basis. Conventional methods of vaccine development are as follows.

LIVE ATTENUATED VACCINES

Live attenuated vaccines are produced by culturing viruses in a laboratory-grown host cell. The virus is administered to the natural host after several passages under different circumstances to ensure that random mutation has generated a non-virulent and replicative infectious agent (Meeusen et al. 2007). Since the virus is in its replicative form and closely mimics the natural infection, these vaccines provide enhanced immunity and eliminate the requirement for repeated boosters. There are several disadvantages associated with attenuated vaccines. The major disadvantage is that these attenuated strains can mutate and revert to a virulent form. The live attenuated polio vaccine developed by Albert Sabin had a reversion rate ranging from 1 case per 1.4 million doses to 1 in every 2.4

Viral Vaccines

million doses (Verdijk, Rots, and Bakker 2011). Live attenuated vaccines can result in other complications and symptoms like other diseases. Sometimes these symptoms can be life-threatening in immunocompromised patients. For example, the MMR vaccination against measles, mumps, and rubella can result in severe neuroinflammation (Poyhonen et al. 2019). With the use of modern genetic engineering tools, virulent genes from the attenuated virus can be mutated or removed permanently, making these vaccines safer. Sometimes these vaccines can suppress the immune system transiently, making vaccinated individuals susceptible to other infections. These vaccines require proper cold storage facilities, and maintaining such facilities is difficult for many developing countries (Yadav, Yadav, and Khurana 2014).

INACTIVATED OR KILLED VACCINES

Inactivated or killed, vaccines are developed by heat or chemical treatment of virulent virus strains. Such treatment makes the virus non-replicative, keeping the structural integrity of the epitopes on antigens intact. Heat treatment is often not used for inactivation as it can result in extensive denaturation of antigenic proteins resulting in a change in the structure of epitopes. Thus, chemical treatment is preferred. For example, the Salk polio vaccine was developed by treating the polio virus with formaldehyde. Several alkylating agents are also used to inactivate the virus. The most used alkylating agent is β-propiolactone (BPL). The vaccine against influenza was developed using BPL (Sanders, Koldijk, and Schuitemaker 2015). Since the virus is in non-replicative form, it requires repeated booster doses to generate a complete immune response. Inactivated vaccines are safer than live attenuated vaccines as Inactivated viruses cannot revert to a virulent form.

NEXT-GENERATION VACCINE DEVELOPMENT

Genomic investigations and a better understanding of pathogenesis have resulted in the discovery of novel antigens and the creation of recombinant vaccines (Pizza et al. 2000). Genome sequencing technologies and approaches used to screen a pathogen's genome and proteome have greatly improved the efficiency of antigen discovery because relevant antigenic structures can be identified and recombinant vaccines can be produced that contain only the antigen required to elicit protective immunity (Seib, Zhao, and Rappuoli 2012). Genomic databases often contain whole genome sequences as well as the entire range of encoded proteins from which vaccine screening is possible (Bagnoli et al. 2011).

WHY NEXT-GENERATION VACCINES ARE REQUIRED

The first-generation or conventional vaccines include live attenuated and inactivated vaccines. These vaccines have provided humans and animals with advantages over the pathogenic world around them. Because of their effectiveness in the animal industry, these vaccinations have had an economic influence (Meeusen et al. 2007). However, certain limitations arise with these conventional vaccines. Both cellular and humoral immune responses are elicited by live attenuated vaccines, which contribute to their potency (da Costa, Walker, and Bonavia 2015; Rizzi et al. 2012). Yet, the possibility for the microbe to revert to a virulent phenotype is a key concern associated with this kind of vaccine. Recombinant vaccines are an appealing technique for overcoming the constraints of conventional vaccines, and several rationally designed and subunit vaccines have already entered the veterinary market (Jorge and Dellagostin 2017).

SECOND-GENERATION VACCINES

Subunit elements, conjugated/recombinant antigens or synthesized proteins are examples of second-generation vaccines.

Recombinant Subunit Vaccines

Recombinant subunit vaccinations do not employ viruses (either inactivated or live) but rather antigen synthesis via antigen overexpression and purification (Aida et al. 2021). Subunit vaccines contain pathogen-specific proteins that are non-infectious. Gene-encoding antigens are frequently cloned and expressed in heterologous systems (Simionatto et al. 2010). Because a virulent or somewhat virulent microorganism is not required to generate immunity, the safety of both the manufacturer and the user is enhanced. Subunit vaccines also have the advantage of including proteins in their most natural state, which aids in proper protein folding and the reconstruction of the conformational epitope (Eshghi et al. 2009). For example, subunit vaccines of HIV are based on envelope proteins (Cohen and Dolin 2013).

Subunit vaccines containing virus-like particles (VLPs) do not include any genetic material capable of replication, but they do present antigens in a repeated, ordered array, analogous to the virion structure, which is thought to boost immunogenicity (Jennings and Bachmann 2008). VLPs can induce humoral and cell-mediated immune responses without adjuvants due to their molecular scaffolds and lack of genomes, which resemble native viruses. However, no commercial vaccination has used these methods (Liu et al. 2012). VLPs have been used as a vaccine for rotavirus and are currently under evaluation for human use (Conner et al. 1996).

Vector Vaccines

The development of novel prophylactic and therapeutic vaccine candidates has been aided using antigen/gene delivery systems. A vector is used in vector vaccine technology to deliver protective protein(s) to the immune system of the vaccinated host. These vectors are typically immunogenic, displaying numerous antigens. There are two types of recombinant vector vaccines: live vector vaccines and naked DNA vaccines (Jorge and Dellagostin 2017).

Live Vector Vaccines

Attenuated bacteria or viruses are known as classical live vectors, and they can be utilized to produce immunogenic antigens from other pathogens in addition to eliciting their innate immunity. As a result, they protect against more than one infection. This can be accomplished by inserting one or more genes encoding protective antigens from different pathogens into the attenuated vaccination strain's genome. Live attenuated viral vectors include herpesviruses and adenoviruses. Poxviruses like the canarypox and vaccinia viruses have also been well-examined as vectors (Shams 2005). Live attenuated chimeric recombinant technology is another novel strategy. This strategy involves exchanging some attenuated vaccine strain genes for their corresponding virulent counterparts.

DNA Vaccines

A DNA vaccine is a plasmid containing a gene from a virus, bacteria or parasite that can express an immunogenic protein in mammalian cells. A plasmid contains the gene of interest, strong eukaryotic promoters for transcriptional control, a polyadenylation signal sequence for persistent and effective translation, and a bacterial origin of replication. Transfected host cells transcribe the plasmid into mRNA and translate it into an antigenic protein. The host immune system detects expressed proteins as foreign, triggering cellular and humoral immune responses (Robinson and Torres 1997). No DNA vaccines are licensed for use in humans in the United States as of 2021. The approach has yet to be demonstrated effective in people, and few experiments have elicited a response strong enough to guard against viral diseases. However, a veterinary DNA vaccine against West Nile virus has been approved for horses. In August 2021, ZyCoV-D was given urgent clearance by Indian regulators. The vaccine was created by Cadila Healthcare and was the first DNA vaccine to be approved for human use against severe acute respiratory syndrome coronavirus-2 (SARS-CoV-2) (Blakney and Bekker 2022).

RNA VACCINES

In 1987, Robert Malone mixed messenger RNA with lipid droplets and introduced it into human cells. Later, he observed that messenger RNA started expressing inside human cells. This was a landmark experiment in the history of RNA vaccines (Malone, Felgner, and Verma 1989). Generally, two types of RNAs—non-replicating mRNA and self-replicating viral RNA—are used in the development of RNA-based vaccines. Conventional mRNA-based vaccines encode only the target antigen, while self-replicating viral RNA not only encodes the target antigen but also encodes viral replication machinery that synthesizes multiple copies of RNA inside the host cell. This enhances the production of antigen (Pardi et al. 2018). Although RNA vaccines offer several advantages over other vaccine approaches, their use was restricted because of inefficient RNA delivery systems until recent advancements in technology by Pieter Cullis, a biochemist at the University of British Columbia in Vancouver, Canada. He used lipid nanoparticle technology for the delivery of messenger RNA. Recently, Moderna, a Cambridge-based company, and Pfizer, a New York-based company, developed RNA-based vaccines for SARS-CoV-2 (Dolgin 2021).

REVERSE VACCINOLOGY

The reverse vaccinology (RV) strategy was first described in 2000 as "making use of the pathogen's genetic sequence" (Rappuoli 2000). Whole genome sequencing projects have generated an immense amount of information. This information, together with advanced bioinformatic tools, has given rise to a new era of vaccine development, known as the third generation of vaccines. One approach is reverse vaccinology. This approach uses the pathogen's genomic sequence to probe antigens that can be potential vaccine candidates, thereby reducing the time of vaccine development by one to two years. With this approach, we can not only study antigens used in conventional methods but also identify new antigens, leading to the discovery of unique mechanisms of immune intervention (Kanampalliwar 2020). The flowchart in Figure 21.1 represents the sequential steps that are followed for designing vaccines using the reverse vaccinology approach.

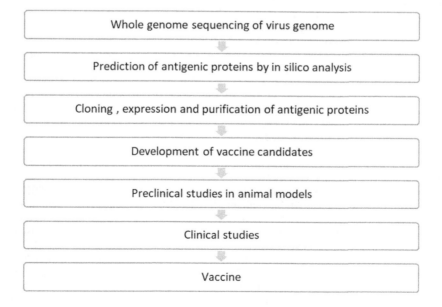

FIGURE 21.1 Flowchart showing the steps involved in the reverse vaccinology approach.

REFERENCES

Aida, V., V. C. Pliasas, P. J. Neasham, J. F. North, K. L. McWhorter, S. R. Glover, and C. S. Kyriakis. 2021. "Novel vaccine technologies in veterinary medicine: a herald to human medicine vaccines." *Front Vet Sci* 8:654289. doi: 10.3389/fvets.2021.654289.

Bagnoli, F., B. Baudner, R. P. Mishra, E. Bartolini, L. Fiaschi, P. Mariotti, V. Nardi-Dei, P. Boucher, and R. Rappuoli. 2011. "Designing the next generation of vaccines for global public health." *OMICS* 15 (9):545–66. doi: 10.1089/omi.2010.0127.

Blakney, A. K., and L. G. Bekker. 2022. "DNA vaccines join the fight against COVID-19." *Lancet* 399 (10332):1281–2. doi: 10.1016/S0140-6736(22)00524-4.

Blume, S., and I. Geesink. 2000. "A brief history of polio vaccines." *Science* 288 (5471):1593–4. doi: 10.1126/science.288.5471.1593.

Bordenave, Guy. 2003. "Louis Pasteur (1822–1895)." *Microbes and Infection* 5 (6):553–60. doi: https://doi.org/10.1016/S1286-4579(03)00075-3.

Coffin, C. S. 2022. "The remarkable success of the vaccine for a killer virus: Hepatitis B." *Hepatology* 75 (6):1365–7. doi: 10.1002/hep.32346.

Cohen, Y. Z., and R. Dolin. 2013. "Novel HIV vaccine strategies: overview and perspective." *Ther Adv Vaccines* 1 (3):99–112. doi: 10.1177/2051013613494535.

Conner, M. E., C. D. Zarley, B. Hu, S. Parsons, D. Drabinski, S. Greiner, R. Smith, B. Jiang, B. Corsaro, V. Barniak, H. P. Madore, S. Crawford, and M. K. Estes. 1996. "Virus-like particles as a rotavirus subunit vaccine." *The Journal of Infectious Diseases* 174 (Supplement 1):S88–S92. doi: 10.1093/infdis/174.Supplement_1.S88.

da Costa, Christopher, Barry Walker, and Aurelio Bonavia. 2015. "Tuberculosis Vaccines – state of the art, and novel approaches to vaccine development." *International Journal of Infectious Diseases* 32:5–12. doi: https://doi.org/10.1016/j.ijid.2014.11.026.

Dolgin, E. 2021. "The tangled history of mRNA vaccines." *Nature* 597 (7876):318–24. doi: 10.1038/d41586-021-02483-w.

Eshghi, A., P. A. Cullen, L. Cowen, R. L. Zuerner, and C. E. Cameron. 2009. "Global proteome analysis of Leptospira interrogans." *J Proteome Res* 8 (10):4564–78. doi: 10.1021/pr9004597.

Francis, T., J. E. Salk, and J. J. Quilligan. 1947. "Experience with vaccination against influenza in the spring of 1947: A preliminary report." *Am J Public Health Nations Health* 37 (8):1013–6. doi: 10.2105/ajph.37.8.1013.

Frierson, J. G. 2010. "The yellow fever vaccine: a history." *Yale J Biol Med* 83 (2):77–85.

Jennings, G. T., and M. F. Bachmann. 2008. "The coming of age of virus-like particle vaccines." *Biol Chem* 389 (5):521–36. doi: 10.1515/bc.2008.064.

Jorge, Sérgio, and Odir Antônio Dellagostin. 2017. "The development of veterinary vaccines: a review of traditional methods and modern biotechnology approaches." *Biotechnology Research and Innovation* 1 (1):6–13. doi: https://doi.org/10.1016/j.biori.2017.10.001.

Kanampalliwar, Amol M. 2020. "Reverse vaccinology and its applications." In *Immunoinformatics*, edited by Namrata Tomar, 1–16. New York, NY: Springer US.

Liu, F., S. Ge, L. Li, X. Wu, Z. Liu, and Z. Wang. 2012. "Virus-like particles: potential veterinary vaccine immunogens." *Res Vet Sci* 93 (2):553–9. doi: 10.1016/j.rvsc.2011.10.018.

Ljubojevic, S. 2006. "The human papillomavirus vaccines." *Acta Dermatovenerol Croat* 14 (3):208.

Malone, R. W., P. L. Felgner, and I. M. Verma. 1989. "Cationic liposome-mediated RNA transfection." *Proc Natl Acad Sci U S A* 86 (16):6077–81. doi: 10.1073/pnas.86.16.6077.

Meeusen, E. N., J. Walker, A. Peters, P. P. Pastoret, and G. Jungersen. 2007. "Current status of veterinary vaccines." *Clin Microbiol Rev* 20 (3):489–510, table of contents. doi: 10.1128/CMR.00005-07.

Pardi, Norbert, Michael J. Hogan, Frederick W. Porter, and Drew Weissman. 2018. "mRNA vaccines — a new era in vaccinology." *Nature Reviews Drug Discovery* 17 (4):261–79. doi: 10.1038/nrd.2017.243.

Pizza, M., V. Scarlato, V. Masignani, M. M. Giuliani, B. Arico, M. Comanducci, G. T. Jennings, L. Baldi, E. Bartolini, B. Capecchi, C. L. Galeotti, E. Luzzi, R. Manetti, E. Marchetti, M. Mora, S. Nuti, G. Ratti, L. Santini, S. Savino, M. Scarselli, E. Storni, P. Zuo, M. Broeker, E. Hundt, B. Knapp, E. Blair, T. Mason, H. Tettelin, D. W. Hood, A. C. Jeffries, N. J. Saunders, D. M. Granoff, J. C. Venter, E. R. Moxon, G. Grandi, and R. Rappuoli. 2000. "Identification of vaccine candidates against serogroup B meningococcus by whole-genome sequencing." *Science* 287 (5459):1816–20. doi: 10.1126/science.287.5459.1816.

Poyhonen, L., J. Bustamante, J. L. Casanova, E. Jouanguy, and Q. Zhang. 2019. "Life-threatening infections due to live-attenuated vaccines: early manifestations of inborn errors of immunity." *J Clin Immunol* 39 (4):376–90. doi: 10.1007/s10875-019-00642-3.

Rappuoli, R. 2000. "Reverse vaccinology." *Curr Opin Microbiol* 3 (5):445–50. doi: 10.1016/s1369-5274(00)00119-3.

Rizzi, C., M. V. Bianco, F. C. Blanco, M. Soria, M. J. Gravisaco, V. Montenegro, L. Vagnoni, B. Buddle, S. Garbaccio, F. Delgado, K. S. Leal, A. A. Cataldi, O. A. Dellagostin, and F. Bigi. 2012. "Vaccination with a BCG strain overexpressing Ag85B protects cattle against Mycobacterium bovis challenge." *PLoS One* 7 (12):e51396. doi: 10.1371/journal.pone.0051396.

Robinson, H. L., and C. A. Torres. 1997. "DNA vaccines." *Semin Immunol* 9 (5):271–83. doi: 10.1006/smim.1997.0083.

Saleh, A., S. Qamar, A. Tekin, R. Singh, and R. Kashyap. 2021. "Vaccine development throughout history." *Cureus* 13 (7):e16635. doi: 10.7759/cureus.16635.

Sanders, Barbara, Martin Koldijk, and Hanneke Schuitemaker. 2015. "Inactivated viral vaccines." In *Vaccine Analysis: Strategies, Principles, and Control*, edited by Brian K. Nunnally, Vincent E. Turula and Robert D. Sitrin, 45–80. Berlin, Heidelberg: Springer.

Seib, K. L., X. Zhao, and R. Rappuoli. 2012. "Developing vaccines in the era of genomics: a decade of reverse vaccinology." *Clin Microbiol Infect* 18 (Supplement 5):109–16. doi: 10.1111/j.1469-0691.2012.03939.x.

Shams, H. 2005. "Recent developments in veterinary vaccinology." *Vet J* 170 (3):289–99. doi: 10.1016/j.tvjl.2004.07.004.

Simionatto, S., S. B. Marchioro, V. Galli, D. D. Hartwig, R. M. Carlessi, F. M. Munari, J. P. Laurino, F. R. Conceicao, and O. A. Dellagostin. 2010. "Cloning and purification of recombinant proteins of Mycoplasma hyopneumoniae expressed in Escherichia coli." *Protein Expr Purif* 69 (2):132–6. doi: 10.1016/j.pep.2009.09.001.

Verdijk, P., N. Y. Rots, and W. A. Bakker. 2011. "Clinical development of a novel inactivated poliomyelitis vaccine based on attenuated Sabin poliovirus strains." *Expert Rev Vaccines* 10 (5):635–44. doi: 10.1586/erv.11.51.

Yadav, Dinesh K., Neelam Yadav, and Satyendra Mohan Paul Khurana. 2014. "Chapter 26 - Vaccines: Present Status and Applications." In *Animal Biotechnology*, edited by Ashish S. Verma and Anchal Singh, 491–508. San Diego: Academic Press.

Index

A

Acquired immunodeficiency syndrome (AIDS), 91
    drugs for advanced-stage, 101, 102
    opportunistic infections in, 98
Adeno-associated virus (AAV), 63–64
Adenoviridae, 56
Adenovirus (AdV), 6, 7
    in cancer treatment, 146, 148–149, 151
    as vaccine vectors, 7
    virions, 56
African swine fever virus (ASFV)
    diagnosis, 160
    etiology and natural reservoirs, 159
    geographical distribution, 159
    history, 159
    isolation, 160
    pathogenesis, 160
    prevention and control, 160
    replication dynamics, 159
    risk factors, 159–160
Agar gel immunodiffusion (AGID), 11, 12
Agar gel precipitin (AGP) test, *see* Agar gel
        immunodiffusion
Agnoprotein, 52
Aichi virus, 75
Allard, Henry A., 140
Alloherpesviridae, 51
Alphacoronavirus, 76
Alphairidovirinae, 53
Anelloviridae, 57, 64
Anelloviruses, 64, 66
Angiotensin-converting enzyme-2 (ACE2), 44
Anthropozoonosis, 17
Antibody-dependent cellular cytotoxicity, 179
Antigenic hypervariability, 179
Antigen-presenting cells, 176
Anti-retroviral therapy (ART), 99–100
Aphids, 18
Aphthovirus, 75
APOBEC proteins, 103
Aquabirnavirus, 72
Aquareovirus, 69
Arboviruses, 17
Arenaviridae, 79
Arenaviruses, 80–81
Arteriviridae, 74, 76
Asfarviridae, 53, 54, 159
Astroviridae, 74
Atadenovirus, 56
Aviadenovirus, 56
Avibirnavirus, 72
Avihepadnavirus, 107
Avipoxvirus, 47

B

Bacilladnaviridae, 57, 62
Bacillus subtilis, 119
Bacterial and Archaeal Viruses Subcommittee (BAVS),
    119
Bacteriophages, 3, 4
    application, 128–129
    bacteria eaters, 8
    classification, 119–120
    complex structures, 33
    definition, 8
    for diagnosis
        phage amplification assays, 130
        phage capture, 130
        phage display system, 129–130
        reporter phage, 130
        utilization, 130–131
    foodborne illness, 8
    history, 119
    life cycle
        lysogenic cycle, 125–127
        lytic cycle, 123–125, 127
        phage particles, 122
    limitations, 8
    phage products, 8
    ssDNA virus integration, 63
    structure, 121–122
Baltimore, David, 26, 91
Baltimore classification (BCII), 57, 69, 78
    Class I, 26
    Class IA, 26
    Class IB, 26
    Class II, 27
    Class III, 27
    Class IV, 27
    Class V, 27
    Class VI, 27–28
    Class VII, 28
    diagrammatic representation, 26, 27
    reverse-transcribing viruses, 104–106
Bang, O., 3, 91
Bawden, Frederick, 137
B-cell receptors, 176
B cells, 176–178
Beak and feather disease virus (BFDV), 66
Beale, Helen Purdy, 140
Beard, John, 146
Beijerinck, Martinus, 1, 140
Betacoronavirus (betaCoV), 76, 164
Betairidovirinae, 53
Bidnaviridae, 57, 60–62
Biogeochemical cycle, 7
Birnaviridae, 72–73

189

Index

Birnavirus, 72, 73
Blubervirales, 106, 107
Bornaviridae, 79
Brenner, Sydney, 2
Bromoviridae, 140
Bunyaviridae, 79

C

Caliciviridae, 74
Campylobacter jejuni, 8
Capripoxvirus, 160, 161
Capsid proteins (CPs), 59, 62
Capsids, 25, 26
    ASFV, 53
    helical, 139
    icosahedral, 139
Capsomeres, 32
Caudovirales, 120, 121
Caudoviricetes, 120
Cauliimoviridae, 104, 106
CD4 T cells, 176, 177
CD8 cytotoxic T lymphocytes (CTLs), 176, 177
Cell-mediated immunity (CML), 176
Chicken anaemia virus (CAV), 66
Chloriridovirus, 53, 54
Chordopoxvirinae (vertebrates), 47, 49
Chronic hepatitis B (CHB), 108, 115
Circoviridae, 62, 64, 65
Circoviruses, 64
Circular Rep-encoding single-stranded (CRESS)
        DNA viruses
    definition, 57
    eukaryotic CRESS DNA viruses, 62
Circulative transmission, 17
Classical live vectors, 184
Classification, virus
    Baltimore system of, 26–28
    capsid structure based, 25, 26
    characteristics, 24
    genome structure and core based, 24, 25
    Hershey–Chase experiment, 24
    principles, 24
Contagium vivum fluidum, 1
Coronaviridae, 75–76, 164
Coronavirinae, 76
Coronaviruses, 76
Covalently closed circular DNA (cccDNA), 109–111
COVID-19, 4, 164–166
Cressdnaviricota, 62
Crick, Francis, 30
Cucumber mosaic virus (CMV), 140–141
Cucumovirus, 140
Cullis, Pieter, 185
Cyclovirus, 64
Cytokines, 178–180

D

Dane particle, 108
DDE transposases, 63
Decapodiridovirus, 53, 54
Deltacoronavirus, 76
d'Herelle, Felix, 1, 8, 119

Diagnostic virology
    electron microscopy, 14
    immunological response detection, 11–13
    light microscopy, 13–14
    molecular methods, 15–16
    virus isolation, 15
DNA vaccines, 184
Double-stranded DNA (dsDNA) viruses
    families
        Adenoviridae, 56
        Asfarviridae, 53, 54
        Herpesviridae, 51
        Iridoviridae, 53–55
        Papillomaviridae, 51–52
        Polyomaviridae, 52–53
        Poxviridae, 47–50
        and structure, 47, 49
        universal taxonomy, 47
    replication properties of, 47
Double-stranded RNA (dsRNA) viruses
    Birnaviridae, 72–73
    polyphyletic and transcribe positive-strand mRNA, 69
    Reoviridae, 69–72
    replication properties, 69
    universal taxonomy of, 69, 70
Dulbecco, Renato, 91
Duplornaviricota, 69

E

Ebola virus, 82
Ectromelia (mousepox) virus, 1
Ellerman, V., 3, 91
Emerging disease, *see* Emerging infectious diseases
Emerging infectious diseases (EIDs)
    African swine fever virus (ASFV), 159–160
    consequences of, 158
    definition, 154
    diagnostic techniques, 156
    FOA and WOAH list, 154, 155
    history, 154–156
    impact of, 154–155
    incidence and global spread, 157–158
    lumpy skin disease virus (LSDV), 160–162
    monkeypox (MPX) virus, 166–167
    Nipah virus (NiV), 162–164
    prevention and control, 158
    risk factors, 156–157
    SARS-CoV-2, 164–166
Emtricitabine (Truvada), 102
Encapsidation, 72, 114
Endogenous retroviruses, 91
Endogenous viral elements (EGEs), 62–63
Enfuvirtide, 100
Enterovirus, 74
Entomobirnavirus, 72
Entomopoxvirinae (insects), 47, 49
Enveloped viruses, 40
Envelope/surface proteins, 113
Enzyme-linked immunosorbent assay (ELISA), 12, 13
Episomes, 125
Epstein–Barr virus, 49, 179
Escherichia coli, 8, 121–124
Eukaryotic viruses, 61

Index

191

Excisionase, 126
Exotic diseases, 154
Extracellular spread, 41

F

Fermentation, 1
Filamentous bacteriophages, 120
Filoviridae, 79
Filoviruses, 81–82
First-generation or conventional vaccines, 183
Flaviviridae, 75–76
Flaviviruses, 75–76
Fluorescence-based methods, 156
Foot-and-mouth disease (FMD), 3, 74
Fraenkel-Conrat, H., 137
Francis, Thomas, Jr., 182
Frosch, Paul, 3, 74
Furovirus, 18

G

Gallo, Robert, 91
Gamaleya, Nikolay Fyodorovich, 119
Gammacoronavirus, 76
Gastrointestinal viruses, 46
Gay-related immunodeficiency (GRID), *see* Acquired
     immunodeficiency syndrome
GB viruses, *see* Pegiviruses
Geminiviridae, 62, 64
Geminiviruses, 33
Gene delivery systems, 6
Gene therapy, 6, 101
Gene transfer, 8
Genome sequencing, 74, 183, 185
Genomoviridae, 62
Global Rinderpest Eradication Programme (GREP), 3
Glycoprotein-cell-associated preprotein (GPC), 80
Glycoproteins (G), 40, 79
     in arenaviruses, 80
     env gene, 92
Goldberg, Tony, 6
Gumboro disease, 72

H

Haagen, Eugen, 182
Hankin, Ernest, 119
Hemagglutination assay (HA), 12
Hemagglutination inhibition (HI), 12
Hendra virus (HeV), 4
Henipaviruses, 3, 162
Hepacivirus, 75–76
Hepadnaviridae, 104
     generas, 106–107
     HBV, *see* Hepatitis B virus
Hepatitis B virus (HBV)
     clinical symptoms, 107
     diagnosis, 115
     epidemiology, 108
     genotypes, 115
     Metahepadnavirus, 106
     molecular biology
          envelope/surface proteins, 113

life cycle, 109–110
     polymerase (P) protein, 112–113
     precore/core protein, 111–112
     replication, 114–115
     transcription and translation, 110–111
     virion morphology and genomic organization,
          108–109
     X protein, 113
Parahepadnavirus, 106
transmission, 107
treatment, 115–116
vaccines, 116
Hepeviridae, 74
Herpes simplex virus (HSV), 51
Herpes simplex virus type 1 (HSV-1), 7, 146, 149, 151
Herpesviridae, 51
Herpesvirus, 6
Herpetohepadnavirus, 106
Heterologous virus immunity, 179
Highly active anti-retroviral therapy (HAART), 99
His hydrophobic His (HUH) endonuclease, 61
History of virology
     bacteriophage, 3
     bat-borne viruses, 3–4
     beneficial properties, 4
     cancer and virus relationship, 3
     cell culture techniques, 1, 3
     conceptual and technological inventions, 1, 2
     influenza virus, 3
     molecular era, 3
     nucleic acid sequences, cloning of, 3
     plant viruses, 1
     polymerase chain reaction (PCR), 3
     rinderpest, 3
     role of electron microscopy, 1–2, 14, 34
     smallpox, 3
Holmes, Francis O., 140
Horizontal transmission, 17, 19
Horne, Robert, 2, 24
Human ACE2 protein, 44
Human endogenous retroviruses (HERVs), 102–103
Human immunodeficiency virus (HIV), 91
     CD4+ cells, 95
     cell-mediated immunity, 95
     clinical findings, 97–98
     cytotoxic T cells, 95
     dendritic cells infection, 94
     epidemiology and transmission, 94, 96
     genome and replication
          envelope proteins, 94
          env gene, 92
          gag and pol polyproteins, 94
          gag gene, 92
          genes and proteins, functions of, 92, 93
          pol genes, 92
          polyproteins, 94
          replication cycle, 95
          structure, 92, 94
          Tat and Nef proteins, 94
          Vif protein, 94
     immune response, 95
     Kaposi's sarcoma (HHV-8), 95
     laboratory diagnosis, 98–99
     Langerhans cells, 94

192 Index

pathogenesis, 95, 97
prevention, 102
structural properties, 91–93
T$_h$ cells, 95
treatment
    drug resistance, 99–100
    entry and fusion inhibitors, 99, 100
    genotyping method, 100
    immune reconstruction inflammatory syndrome
        (IRES), 101, 102
    integrase inhibitors, 101
    nonnucleoside RT inhibitors (NNRTIs), 101
    nucleoside/nucleotide RT inhibitors (NRTIs), 101
    phenotyping method, 100
    protease inhibitors, 101
    vaccines, 101
Human rhinoviruses, 75
Human T-cell lymphotropic virus 1 (HTLV-1), 91, 102
Humoral immunity, 176

I

Ichtadenovirus, 56
IgA antibody, 177
IgD antibody, 177
IgG antibody, 177
IgM antibody, 177
Immune reconstruction inflammatory syndrome (IRES),
    101, 102
Immunity, viral infection
    adaptive, 176–177
    autoimmunity, 179–180
    host, 180
    immune evasion, 178–179
    immunological memory, 178
    immunopathology, 179
    inborn errors of immunity (IEIs), 178
    innate, 174–176
    vaccine development, 180
Immunohistochemistry, 163
Immunosuppression, 180
Inactivated/killed vaccines, 183
Inborn errors of immunity (IEIs), 178
Incubation period, 43
Influenza virus, 3
    A, B, an C, 82
    HA and HI test, 12
    replication, 39
Inoviridae, 120
Integrative viral vectors, 6
Interferons (IFNs), 176
International Committee on Nomenclature of Viruses
    (ICNV), 138
International Committee on Taxonomy of Viruses (ICTV),
    24, 57, 69, 106, 119–120, 138
Intracellular enveloped virus (IEV), 50
Intracellular mature virus (IMV), 50
Intracellular spread, 41
Iridoviridae, 53–55
Iridovirus, 53, 54
Ivanovsky, Dmitry, 1

J

Jenner, Edward, 47, 180, 182

K

Kaposi's sarcoma (HHV-8), 95
Knoll, Max, 1

L

Landsteiner, Karl, 74
Lentivirinae, 91
Lentiviruses, 4
Listeria monocytogenes, 8
Live attenuated vaccines, 182–183
Loeffler, Friedrich, 3, 74
Loop-mediated isothermal amplification (LAMP), 156
Lopez, Susana, 6
Lowenstein, A., 51
LTAg protein, 52–53
Lumpy skin disease virus (LSDV)
    diagnosis, 162
    geographical distribution, 161
    history, 160–161
    isolation, 162
    pathogenesis, 162
    prevention and control, 162
    properties and host range, 161
    replication dynamics, 161
    risk factors, 161
Lundstrom, K., 150
Lwoff, A., 24
Lymphocystivirus, 53, 54
Lysogenic cycle, bacteriophages
    bacteriophage lambda (λ) DNA genome, 125
    cII and cIII proteins, 125, 126
    cro protein, 127
    decision-making process, 125, 126
    induction, 125
    prophage, 125
    selection, factors influencing, 127
    temperate bacteriophages, 125
    transcription, 125
    λ repressor, 126–127
Lyssavirus, 84
Lytic cycle, bacteriophages
    assembly process, 124
    endolysin, 124
    packasome, 123
    peptidoglycan, 125
    selection, factors influencing, 127
    steps, 123, 124
    synthesis, 123
    T4 virulent phage, 123

M

Malacoherpesviridae, 51
Malone, Robert, 185
Maraba virus, 151
Marchoux, Emile, 155
Mastadenovirus, 56
Matrix protein (M), 79
McKinney, Howard H., 140
Measles infection, 83, 86
Medin, Oskar, 74
Megalocytivirus, 53, 54
Messenger RNA (mRNA), 74
Metagenomics, 61

Index

Metaxyviridae, 62
Meyer, Adolf, 1
Microarray technology, 156
Millman, Irving, 182
Minus-strand RNA viruses, *see* Negative-sense
    RNA viruses
Mollentze, N., 67
Monkeypox (MPX) virus
    classification and genome characteristics, 166
    diagnosis, 167
    epidemiology, 167
    host and transmission pattern, 167
    prevention and control, 167
Mononegavirales, 78, 79
Montgomery, Robert Eustace, 159
mRNA transcripts, 48
Mucosal secretory IgA (sIgA), 177

N

Nackednaviruses, 106
Nanoviridae, 62, 64
Natural resistance-associated macrophage protein
    (NRAMP), 44
Negative-contrast electron microscopy, 2
Negative-sense RNA viruses
    applications, 86
    benefits
        control over gene expression, 79
        reduced risk of recombination, 79
        segmented genomes, 79–80
    families of, 78–84
    genome replication and transcription, 84–85
    nucleocapsid, 78
    structural proteins, 78–79
    viral envelope, 78
Nene, Y. L., 1
Nepoviruses, 18
Nevirapine, 101
Nidovirales, 76
Nipah virus (NiV), 4
    classification and genome characteristics, 162–163
    diagnosis, 163–164
    epidemiology, 163
    host and transmission, 163
    prevention and control, 164
Non-coding control region (NCCR), 52–53
Non-enveloped viruses, 40
Non-integrative viral vectors, 6
Nonnucleoside RT inhibitors (NNRTIs), 101
Nucleocapsid(s), 30
    HBV, 109
    helical symmetry, 31
    herpesvirus replication, 51
    icosahedral symmetry, 32
    negative-sense RNA viruses, 78
    of negative-stranded RNA genomes, 139
Nucleoprotein (NP), 78
Nucleoside/nucleotide RT inhibitors (NRTIs), 101
Nucleos(t)ide analogues (NAs) therapy, 115
Nudnaviridae, 106

O

Okamoto, H., 115
Oncogenic retroviruses, 102

Oncolysis, 146
Oncolytic viruses (OVs), 6–7
    biology and mechanism
        immunity, 148
        tumor cell killing mechanisms, 148
        viral selection and modification, 146–148
    challenges of, 151
    common viruses, 147
    future perspectives, 152
    history, 146
    oncolytic potency measurement, 150
    therapy
        clinical trials in, 151–152
        overcoming barriers, 152
    types
        DNA viruses, 148–150
        RNA viruses, 149–150
Oncorine, 7
Oncornaviruses, 91
Oncovirinae, 91
Oncoviruses, 102
Orbivirus replication, 72
Ortervirales, 106
Orthocoronavirinae, 164
Orthohepadnavirus, 107
Orthomyxoviridae, 79, 82
Orthomyxoviruses, 82, 83
Orthopoxvirus, 47, 166
Orthoreovirus, 69
    σ1 protein of, 71
Orthoretroviruses, 91

P

Packasome, 123
Papillomaviridae, 51–52
Paramyxoviridae, 79, 162
Paramyxovirinae, 82
Paramyxoviruses, 31, 82–83
Parapoxvirus, 47
Parvoviridae, 57, 60–62, 64, 65
Parvovirus endonuclease, 63–64
Parvoviruses, 64
Pasteur, Louis, 1, 47, 182
Pathogenesis
    acute and persistent infections, 44–45
    host restriction factors, 45–46
    incubation period, 43
    local viral dissemination and replication, 41–42
    overview of, 41
    proviral factors, 45
    receptor interactions, 43–44
    shedding of virus, 46
    states of infections, 44
    tissue tropism, 43
    viral spread
        in bloodstream, 42
        conjunctivitis epidemic, 42
        extracellular and intracellular, 41
        in nerve, 42
    virus implantation, 41
Pathogen recognition receptors, 174
Pegiviruses, 75–76
Pestiviruses, 75–76
Phages, *see* Bacteriophages
Phlebovirus, 156

194

Photoluminescence (PL), 156
Picobirnaviruses, 72
Picornaviridae, 74–75
Picornaviruses, 74–75
Pirie, Norman, 137
Pisuviricota, 69
Plant viruses, 1, 7
    applications in biotechnology, 141–143
    classification, 138
    cucumber mosaic virus (CMV), 140–141
    diversity, 138
    host–virus interactions, 139–140
    Monodnaviria, 138
    Riboviria, 138
    RNAi system, 137
    structure of, 138–139
    tobacco mosaic virus (TMV), 137, 140
    transmission
        modes of, 139
        vectors, 18, 142
Plant virus nanoparticles (PVNPs), 143
Plasma membrane, direct fusion, 37
Pleolipoviridae, 57, 63
Pleomorphic viruses, 120
Pneumovirinae, 82
Poliovirus, 74, 183
Polymerase and phosphoprotein (L and P), 78
Polymerase chain reaction (PCR), 3, 15–16
    EID/TAD diagnostics, 156
Polymerase (P) protein, 112–113
Polyomaviridae, 52–53
Popper, Emil, 74
Porcine circovirus 2 (PCV2), 66
Positive-sense RNA viruses
    families
        Coronaviridae, 74–76
        Flaviviridae, 75–76
        Picornaviridae, 74–75
    properties, 74
Postexposure prophylaxis (PEP), 102
Poxviridae, 47–50, 160, 161, 166
Poxviruses, 33
    replication, 38, 39
    vaccine, 184
Precore/core ORF, 111–112
Preexposure prophylaxis (PrEP), 102
Pregenomic RNA (pgRNA), 110, 111, 114
Prokaryotic viruses, 61
Prophages, 125
Protein-primed DNA polymerase-mediated replication,
    59–61
Pseudolysogeny, 123
Pteropodidae, 163
Pteropus, 163

Q

Quantum dots (QDs), 156

R

Rabies virus (RABV), 84
Ranavirus, 53, 54
Receptor-mediated endocytosis, 37–38

Recombinant subunit vaccines, 184
Redondoviridae, 62, 64
Re-emerging disease, 154
Reoviridae/reoviruses
    in cancer treatment, 149
    classification and properties, 69–70
    genome organization and replication, 70–72
    history, 69
Replication, virus
    Adenoviridae, 56
    of African swine fever virus, 159
    arenaviruses, 80–81
    Asfarviridae, 53, 54
    components of cycle
        attachment, 35–37
        genome replication, 38–40
        viral assembly and release, 40
        viral entry into host cell, 37–38
    eclipse period, 35, 36
    HBV, 114–115
    Herpesviridae, 51
    Iridoviridae, 54–55
    lumpy skin disease virus, 161
    negative-sense RNA viruses, 84–85
    oncolytic viruses, 147
    one-step growth curve, 35, 36
    overview of, 35, 36
    Papillomaviridae, 51–52
    Polyomaviridae, 52–53
    Poxviridae, 47–50
    properties, dsRNA viruses, 69
    in ssDNA viruses, 59–61
Replication-associated protein (Rep), 57, 59
Respiratory viruses, 46
Retroviridae, 106
Retroviruses, 3, 40
    chicken sarcoma, 91
    classification, 91, 92
    discovery of, 91
    endogenous, 102–103
    HIV, *see* Human immunodeficiency virus
    oncogenic, 102
    structural properties, 91–93
Reverse-transcribing DNA viruses
    classification, 104–106
    families, 104, 106–107
    HBV, *see* Hepatitis B virus
    taxonomy, 106
Reverse-transcribing viruses, 104
    central dogma of molecular biology, 104, 105
Reverse transcriptase (RT), 91
Reverse transcriptase PCR (RT-PCR), 15
Reverse vaccinology (RV), 185
Reverse zoonosis, *see* Zooanthroponosis
Revtraviricetes, 104, 106
Rhabdoviridae, 79
Rhabdoviruses, 31, 83–84, 86
Ribonucleoprotein (RNP), 80
Riboviria, 107
Rift Valley fever (RVF), 156
Rigvir, 7, 151
RNA-dependent DNA polymerase (RdDp), 91, 104
RNA-dependent RNA polymerase (RdRp), 69, 78, 84–85
RNA helicase family, 174

Index

RNA interference (RNAi), 137
RNA vaccines, 185
Rolling circle replication (RCR), 59, 60
Rolling hairpin replication (RHR), 59–61
Roniviridae, 76
Rotavirus
    receptor-dependent endocytosis, 72
    replication, 72
    structure of, 70
    VLPs as vaccine, 184
    VP4 and VP7, 72
Rotavirus B, 69
Rous, Peyton, 3, 91
Rous sarcoma virus (RSV), 3, 91
Ruska, Ernst, 1

S

Sabin, Albert, 182
Salk, Jonas, 182
Salmonella spp., 8
Sandwich enzyme-linked immunosorbent assay (sELISA), 156
SARS coronavirus (SARS-CoV), 3, 4, 44
Second-generation vaccines, 183
Sedoreovirinae, 70, 71
Serine/threonine integrases, 63
Severe acute respiratory syndrome coronavirus 2 (SARS-CoV-2)
    classification and genome characteristics, 164–165
    diagnosis, 165–166
    epidemiology, 165
    host and transmission pattern, 165
    prevention and control, 166
Sexual transmission, 17, 22
Sharma, G.S., 3
Shigella spp., 8
Siadenovirus, 56
Single-stranded DNA (ssDNA) viruses
    eukaryotic CRESS DNA viruses, 62
    genome architecture, 57, 59
    integration into host genomes
        into archaeal genomes, 63
        in bacterial genomes, 63
        endogenous viral elements (EGEs), 62–63
        into eukaryotic genomes, 63–64
    origin and evolution, 61–62
    pathogenic, 64–66
    taxonomy and classification, 57, 58
    transboundary movement and zoonotic potential, 66–67
    virus replication, 59–61
Single-stranded RNA viruses, 150
    Nipah virus, 162
Smacoviridae, 62, 64
Smallpox, 155
Smith, Hugh, 182
Someshvardeva, 1
Southam, Chester, 146
Spike protein (S), 44
Spinareovirinae, 70, 71
Spiraviridae, 63
Spumaretrovirinae, 106
Spumavirinae, 91

Spumaviruses, 91
ssDNA circular genome integration, 64
STAg protein, 52–53
Stanley, Wendell M., 137
Structure, virus, 29–34; *see also* Virion
Surapala, 1
Systemic IgG, 177

T

Talimogene laherparepvec (T-VEC), 7, 146, 151
T cells, 177, 178
Temin, H., 91
Testadenovirus, 56
Theiler, Max, 182
Tissue tropism, 43
Tobacco disease, 1
Tobacco mosaic virus (TMV), 1, 137, 140
    helical symmetry in, 31
Tobraviruses, 18
Togaviridae, 74
Toll-like receptor (TLR) family, 174
Torovirinae, 76
Torque teno virus (TTV), 65
Tournier, Paul, 24
Transboundary animal diseases (TADs), *see* Emerging infectious diseases
Transmission, viruses
    routes of entry
        accessibility of host cells, 20
        conjunctiva, 22–23
        gastrointestinal entry portal, 22
        genital tract, 22
        permissibility of host cells, 20
        respiratory tract entry portal, 21–22
        skin entry portal, 20–22
        susceptibility of host cells, 20
    by vectors
        arthropods, 17
        plant viruses, 18
    vertebrate viruses, 19
    vertical and horizontal transmission, 17, 19
    zoonotic diseases, 17, 18
Transovarial transmission, 17
TRIM5α molecule, 45
Triple-drug therapy, *see* Anti-retroviral therapy
Trofile assay, 100
Twort, Frederick, 119
Tyrrell, David A.J., 76

V

Vaccines
    development, 180
        next-generation, 183–185
        strategies, 182–183
    hepatitis B virus (HBV), 116
    historical perspective, 182
    human immunodeficiency virus (HIV), 101
    reverse vaccinology (RV), 185
Vaccinia virus, 1, 6, 7
Vector vaccines, 184
Vertebrate viruses
    nonvector transmission, 19

Index

vector transmission, 19
Vertical transmission, 17, 19
Vesicular stomatitis viruses (VSV), 149
Vidal, Jean Baptiste Émile, 51
Viral DNA synthesis, 55
Viral envelope, 78
Viral genome
  delivery, 30
  direct penetration of, 37
  packaging of, 33
  protection, 29–30
  replication, 38–40
    iridovirid, 54–55
    Papillomaviridae, 51–52
  sense genomes, 57
  transcriptional regions, 52
Viral inclusion bodies (VIBs), 71
Viral infection
  causing immune suppression, 180
  host responses, 174, 175
  immunity, *see* Immunity, viral infection
Viral macromolecular synthesis, 54
Viral predation, 7
Viral proteins, 4, 11, 40
  hexon, 56
  icosahedral capsid, 32
  N, P, M, G, and L proteins, 83
  pE248R, 53
  in polyomaviruses, 52
Virion
  of ASFV, 159
  complex virus structure, 33
  definition, 29
  with envelopes, 29, 33
  functions of proteins, 29–30
  lipoprotein membrane, 139
  nomenclature, 30
  primary function of, 29
  protective protein coat, 29, 30
  proteolytic cleavage, 50
  structure in *Asfarviridae*, 53
  study methods, 34
  viral capsids, 29
    with helical symmetry, 31
    with icosahedral symmetry, 32–33
  viral genome packaging, 33
  visualization, 138
Viroplasms, 48; *see also* Viral inclusion bodies

Virotherapy, 6
Virulence levels, 72
Viruses; *see also individual entries*
  adeno-associated viruses, 4
  application
    in agriculture, 7
    in ecology, 7–8
    in food industry, 8
    in medicine, 6–7
  and cancer relationship, 3
  classification, *see* Classification, virus
  etymology, 1
  influenza, 3
  isolation, 15
  life cycle of, 17
  nucleic acids, 3
  particles, *see* Virion
  plant, *see* Plant viruses
  proteins, *see* Viral proteins
  and protists, 7–8
  replication, *see* Replication, virus
  retrovirus, 3, 40
  structure, 29–34
  transmission, *see* Transmission, viruses
Virus factories, *see* Viroplasms
Virus-like particles (VLPs), 184
Virus neutralization (VN) test, 12, 13
von Borries, Bodo, 1
von Heine, Jacob, 74

W

Watson, James, 30
Watt, Sir George, 3

X

XerC/XerD tyrosine recombinase machinery, 63
X protein, 113

Z

Zidovudine, 101
Zooanthroponosis, 17
Zoonotic diseases, 17, 18, 66–67
Zoonotic viruses, 44
  monkeypox, *see* Monkeypox virus
  NiV, *see* Nipah virus